Extrusion problems solved

Related titles:

Extrusion cooking
(ISBN 978-1-85573-559-0)
Extrusion cooking is the ideal method for manufacturing a number of food products from snacks and breakfast cereals to baby foods. However, as a complex multivariate process it requires careful control if product quality is to be maintained. Edited by a leading authority in the field, and with an international team of contributors, this important collection reviews some of the key factors affecting quality and how they can be controlled in manufacturing a range of extruded products. It has been widely welcomed as a key reference in maximising the quality of extruded products.

Food processing technology (Third edition)
(ISBN 978-1-84569-216-2)
The first edition of *Food processing technology* was quickly adopted as the standard text by many food science and technology courses. This completely revised and updated third edition consolidates the position of this textbook as the best single-volume introduction to food manufacturing technologies available. This edition has been updated and extended to include the many developments that have taken place since the second edition was published. In particular, advances in microprocessor control of equipment, 'minimal' processing technologies, functional foods, developments in 'active' or 'intelligent' packaging, and storage and distribution logistics are described. Technologies that relate to cost savings, environmental improvement or enhanced product quality are highlighted. Additionally, sections in each chapter on the impact of processing on foodborne microorganisms are included for the first time.

Thermal technologies in food processing
(ISBN 978-1-85573-558-3)
Thermal technologies have long been at the heart of food processing. The application of heat is both an important method of preserving foods and a means of developing texture, flavour and colour. An essential issue for food manufacturers is the effective application of thermal technologies to achieve these objectives without damaging other desirable sensory and nutritional qualities in a food product. Edited by a leading authority in the field, and with a distinguished international team of contributors, *Thermal technologies in food processing* addresses this major issue. It provides food manufacturers and researchers with an authoritative review of thermal processing and food quality.

Details of these books and a complete list of Woodhead's titles can be obtained by:

- visiting our web site at www.woodheadpublishing.com
- contacting Customer Services (e-mail: sales@woodheadpublishing.com; fax: +44 (0)1223 832819; tel.: +44 (0)1223 499140 ext. 130; address: Woodhead Publishing Limited, 80, High Street, Sawston, Cambridge CB22 3HJ, UK)
- contacting our US office (e-mail: usmarketing@woodheadpublishing.com; tel. (215) 928 9112; address: Woodhead Publishing, 1518 Walnut Street, Suite 1100, Philadelphia, PA 191023-3406, USA)

If you would like to receive information on forthcoming titles, please send your address details to: Francis Dodds (address, tel. and fax as above; e-mail: francis.dodds@woodhead publishing.com). Please confirm which subject areas you are interested in.

Woodhead Publishing Series in Food Science, Technology and Nutrition:
Number 226

Extrusion problems solved

Food, pet food and feed

Mian N. Riaz and Galen J. Rokey

WOODHEAD PUBLISHING

Oxford Cambridge Philadelphia New Delhi

Published by Woodhead Publishing Limited,
80 High Street, Sawston, Cambridge CB22 3HJ, UK
www.woodheadpublishing.com

Woodhead Publishing, 1518 Walnut Street, Suite 1100, Philadelphia, PA 19102-3406, USA

Woodhead Publishing India Private Limited, G-2, Vardaan House, 7/28 Ansari Road,
Daryaganj, New Delhi – 110002, India
www.woodheadpublishingindia.com

First published 2012, Woodhead Publishing Limited

British Library Cataloguing in Publication Data
A catalogue record for this book is available from the British Library.

Library of Congress Control Number: 2011935502

ISBN 978-1-84569-664-1 (print)
ISBN 978-0-85709-520-6 (online)
ISSN 2042-8049 Woodhead Publishing Series in Food Science, Technology and Nutrition (print)
ISSN 2042-8057 Woodhead Publishing Series in Food Science, Technology and Nutrition (online)

The publisher's policy is to use permanent paper from mills that operate a sustainable forestry policy, and which has been manufactured from pulp which is processed using acid-free and elemental chlorine-free practices. Furthermore, the publisher ensures that the text paper and cover board used have met acceptable environmental accreditation standards.

Typeset by Ann Buchan (Typesetters), Middlesex, UK
Printed by TJI Digital, Padstow, Cornwall, UK

Contents

About the authors

Mian Riaz is currently the Director of the Food Protein R&D Center, Head of the Extrusion Technology Program, and a graduate faculty member in the Food Science and Technology Program at Texas A&M University, USA. His research focuses on extrusion of food and feed, extruded snacks, texturization of vegetable protein, utilization of by-products by extrusion, oilseed extrusion and extrusion expelling of oilseed crops. He has published four books, among them *Extruders in food applications* and *Extruders and expanders in pet food, aquatic and livestock feeds*, as well as 16 book chapters and more than 100 papers on extrusion and related topics. He is also a frequent speaker at international conferences and has delivered more than 150 presentations across the world, as well as organizing four courses a year in the area of extrusion at Texas A&M University.

Dr Mian N. Riaz
Director, Food Protein R&D Center
2476 TAMU
Texas A&M University
College Station
Texas 77843 – 2476
USA
E-mail: mnriaz@tamu.edu

Galen Rokey is the Process Manager for the Pet Food Applications Group within Wenger Manufacturing, a major supplier of extrusion and dryer process hardware and technology for the food and feed industries. A graduate of Kansas State University, he has 38 years of laboratory and research experience in the field of extrusion and has authored numerous publications regarding the extrusion process. He is a past member of the American Association of Cereal Chemists and the Institute of Food Technologists and was the recipient of the Alpha Mu Distinguished Service Award in Extrusion from Kansas State University in 1990. He also travels extensively, providing training and assistance in process issues and is a frequent speaker at extrusion short courses at Texas A&M University.

Galen J. Rokey
Process Technology Manager
Wenger Manufacturing, Inc.
714 Main Street
PO Box 130
Sabetha
Kansas 66534 – 0130
USA
E-mail: grokey@wenger.com

Woodhead Publishing Series in Food Science, Technology and Nutrition

1 **Chilled foods: a comprehensive guide** *Edited by C. Dennis and M. Stringer*
2 **Yoghurt: science and technology** *A. Y. Tamime and R. K. Robinson*
3 **Food processing technology: principles and practice** *P. J. Fellows*
4 **Bender's dictionary of nutrition and food technology Sixth edition** *D. A. Bender*
5 **Determination of veterinary residues in food** *Edited by N. T. Crosby*
6 **Food contaminants: sources and surveillance** *Edited by C. Creaser and R. Purchase*
7 **Nitrates and nitrites in food and water** *Edited by M. J. Hill*
8 **Pesticide chemistry and bioscience: the food-environment challenge** *Edited by G. T. Brooks and T. Roberts*
9 **Pesticides: developments, impacts and controls** *Edited by G. A. Best and A. D. Ruthven*
10 **Dietary fibre: chemical and biological aspects** *Edited by D. A. T. Southgate, K. W. Waldron, I. T. Johnson and G. R. Fenwick*
11 **Vitamins and minerals in health and nutrition** *M. Tolonen*
12 **Technology of biscuits, crackers and cookies Second edition** *D. Manley*
13 **Instrumentation and sensors for the food industry** *Edited by E. Kress-Rogers*
14 **Food and cancer prevention: chemical and biological aspects** *Edited by K. W. Waldron, I. T. Johnson and G. R. Fenwick*
15 **Food colloids: proteins, lipids and polysaccharides** *Edited by E. Dickinson and B. Bergenstahl*
16 **Food emulsions and foams** *Edited by E. Dickinson*
17 **Maillard reactions in chemistry, food and health** *Edited by T. P. Labuza, V. Monnier, J. Baynes and J. O'Brien*
18 **The Maillard reaction in foods and medicine** *Edited by J. O'Brien, H. E. Nursten, M. J. Crabbe and J. M. Ames*
19 **Encapsulation and controlled release** *Edited by D. R. Karsa and R. A. Stephenson*
20 **Flavours and fragrances** *Edited by A. D. Swift*
21 **Feta and related cheeses** *Edited by A. Y. Tamime and R. K. Robinson*
22 **Biochemistry of milk products** *Edited by A. T. Andrews and J. R. Varley*
23 **Physical properties of foods and food processing systems** *M. J. Lewis*

24 **Food irradiation: a reference guide** *V. M. Wilkinson and G. Gould*
25 **Kent's technology of cereals: an introduction for students of food science and agriculture Fourth edition** *N. L. Kent and A. D. Evers*
26 **Biosensors for food analysis** *Edited by A. O. Scott*
27 **Separation processes in the food and biotechnology industries: principles and applications** *Edited by A. S. Grandison and M. J. Lewis*
28 **Handbook of indices of food quality and authenticity** *R. S. Singhal, P. K. Kulkarni and D. V. Rege*
29 **Principles and practices for the safe processing of foods** *D. A. Shapton and N. F. Shapton*
30 **Biscuit, cookie and cracker manufacturing manuals Volume 1: ingredients** *D. Manley*
31 **Biscuit, cookie and cracker manufacturing manuals Volume 2: biscuit doughs** *D. Manley*
32 **Biscuit, cookie and cracker manufacturing manuals Volume 3: biscuit dough piece forming** *D. Manley*
33 **Biscuit, cookie and cracker manufacturing manuals Volume 4: baking and cooling of biscuits** *D. Manley*
34 **Biscuit, cookie and cracker manufacturing manuals Volume 5: secondary processing in biscuit manufacturing** *D. Manley*
35 **Biscuit, cookie and cracker manufacturing manuals Volume 6: biscuit packaging and storage** *D. Manley*
36 **Practical dehydration Second edition** *M. Greensmith*
37 **Lawrie's meat science Sixth edition** *R. A. Lawrie*
38 **Yoghurt: science and technology Second edition** *A. Y. Tamime and R. K. Robinson*
39 **New ingredients in food processing: biochemistry and agriculture** *G. Linden and D. Lorient*
40 **Benders' dictionary of nutrition and food technology Seventh edition** *D. A. Bender and A. E. Bender*
41 **Technology of biscuits, crackers and cookies Third edition** *D. Manley*
42 **Food processing technology: principles and practice Second edition** *P. J. Fellows*
43 **Managing frozen foods** *Edited by C. J. Kennedy*
44 **Handbook of hydrocolloids** *Edited by G. O. Phillips and P. A. Williams*
45 **Food labelling** *Edited by J. R. Blanchfield*
46 **Cereal biotechnology** *Edited by P. C. Morris and J. H. Bryce*
47 **Food intolerance and the food industry** *Edited by T. Dean*
48 **The stability and shelf-life of food** *Edited by D. Kilcast and P. Subramaniam*
49 **Functional foods: concept to product** *Edited by G. R. Gibson and C. M. Williams*
50 **Chilled foods: a comprehensive guide Second edition** *Edited by M. Stringer and C. Dennis*
51 **HACCP in the meat industry** *Edited by M. Brown*
52 **Biscuit, cracker and cookie recipes for the food industry** *D. Manley*
53 **Cereals processing technology** *Edited by G. Owens*
54 **Baking problems solved** *S. P. Cauvain and L. S. Young*
55 **Thermal technologies in food processing** *Edited by P. Richardson*
56 **Frying: improving quality** *Edited by J. B. Rossell*
57 **Food chemical safety Volume 1: contaminants** *Edited by D. Watson*
58 **Making the most of HACCP: learning from others' experience** *Edited by T. Mayes and S. Mortimore*
59 **Food process modelling** *Edited by L. M. M. Tijskens, M. L. A. T. M. Hertog and B. M. Nicolaï*
60 **EU food law: a practical guide** *Edited by K. Goodburn*
61 **Extrusion cooking: technologies and applications** *Edited by R. Guy*

103 **Mycotoxins in food: detection and control** *Edited by N. Magan and M. Olsen*
104 **Improving the thermal processing of foods** *Edited by P. Richardson*
105 **Pesticide, veterinary and other residues in food** *Edited by D. Watson*
106 **Starch in food: structure, functions and applications** *Edited by A.-C. Eliasson*
107 **Functional foods, cardiovascular disease and diabetes** *Edited by A. Arnoldi*
108 **Brewing: science and practice** *D. E. Briggs, P. A. Brookes, R. Stevens and C. A. Boulton*
109 **Using cereal science and technology for the benefit of consumers: proceedings of the 12th International ICC Cereal and Bread Congress, 24–26th May, 2004, Harrogate, UK** *Edited by S. P. Cauvain, L. S. Young and S. Salmon*
110 **Improving the safety of fresh meat** *Edited by J. Sofos*
111 **Understanding pathogen behaviour in food: virulence, stress response and resistance** *Edited by M. Griffiths*
112 **The microwave processing of foods** *Edited by H. Schubert and M. Regier*
113 **Food safety control in the poultry industry** *Edited by G. Mead*
114 **Improving the safety of fresh fruit and vegetables** *Edited by W. Jongen*
115 **Food, diet and obesity** *Edited by D. Mela*
116 **Handbook of hygiene control in the food industry** *Edited by H. L. M. Lelieveld, M. A. Mostert and J. Holah*
117 **Detecting allergens in food** *Edited by S. Koppelman and S. Hefle*
118 **Improving the fat content of foods** *Edited by C. Williams and J. Buttriss*
119 **Improving traceability in food processing and distribution** *Edited by I. Smith and A. Furness*
120 **Flavour in food** *Edited by A. Voilley and P. Etievant*
121 **The Chorleywood bread process** *S. P. Cauvain and L. S. Young*
122 **Food spoilage microorganisms** *Edited by C. de W. Blackburn*
123 **Emerging foodborne pathogens** *Edited by Y. Motarjemi and M. Adams*
124 **Benders' dictionary of nutrition and food technology Eighth edition** *D. A. Bender*
125 **Optimising sweet taste in foods** *Edited by W. J. Spillane*
126 **Brewing: new technologies** *Edited by C. Bamforth*
127 **Handbook of herbs and spices Volume 3** *Edited by K. V. Peter*
128 **Lawrie's meat science Seventh edition** *R. A. Lawrie in collaboration with D. A. Ledward*
129 **Modifying lipids for use in food** *Edited by F. Gunstone*
130 **Meat products handbook: practical science and technology** *G. Feiner*
131 **Food consumption and disease risk: consumer-pathogen interactions** *Edited by M. Potter*
132 **Acrylamide and other hazardous compounds in heat-treated foods** *Edited by K. Skog and J. Alexander*
133 **Managing allergens in food** *Edited by C. Mills, H. Wichers and K. Hoffman-Sommergruber*
134 **Microbiological analysis of red meat, poultry and eggs** *Edited by G. Mead*
135 **Maximising the value of marine by-products** *Edited by F. Shahidi*
136 **Chemical migration and food contact materials** *Edited by K. Barnes, R. Sinclair and D. Watson*
137 **Understanding consumers of food products** *Edited by L. Frewer and H. van Trijp*
138 **Reducing salt in foods: practical strategies** *Edited by D. Kilcast and F. Angus*
139 **Modelling microorganisms in food** *Edited by S. Brul, S. Van Gerwen and M. Zwietering*
140 **Tamime and Robinson's Yoghurt: science and technology Third edition** *A. Y. Tamime and R. K. Robinson*
141 **Handbook of waste management and co-product recovery in food processing Volume 1** *Edited by K. W. Waldron*

142 **Improving the flavour of cheese** *Edited by B. Weimer*
143 **Novel food ingredients for weight control** *Edited by C. J. K. Henry*
144 **Consumer-led food product development** *Edited by H. MacFie*
145 **Functional dairy products Volume 2** *Edited by M. Saarela*
146 **Modifying flavour in food** *Edited by A. J. Taylor and J. Hort*
147 **Cheese problems solved** *Edited by P. L. H. McSweeney*
148 **Handbook of organic food safety and quality** *Edited by J. Cooper, C. Leifert and U. Niggli*
149 **Understanding and controlling the microstructure of complex foods** *Edited by D. J. McClements*
150 **Novel enzyme technology for food applications** *Edited by R. Rastall*
151 **Food preservation by pulsed electric fields: from research to application** *Edited by H. L. M. Lelieveld and S. W. H. de Haan*
152 **Technology of functional cereal products** *Edited by B. R. Hamaker*
153 **Case studies in food product development** *Edited by M. Earle and R. Earle*
154 **Delivery and controlled release of bioactives in foods and nutraceuticals** *Edited by N. Garti*
155 **Fruit and vegetable flavour: recent advances and future prospects** *Edited by B. Brückner and S. G. Wyllie*
156 **Food fortification and supplementation: technological, safety and regulatory aspects** *Edited by P. Berry Ottaway*
157 **Improving the health-promoting properties of fruit and vegetable products** *Edited by F. A. Tomás-Barberán and M. I. Gil*
158 **Improving seafood products for the consumer** *Edited by T. Børresen*
159 **In-pack processed foods: improving quality** *Edited by P. Richardson*
160 **Handbook of water and energy management in food processing** *Edited by J. Klemeš, R. Smith and J.-K. Kim*
161 **Environmentally compatible food packaging** *Edited by E. Chiellini*
162 **Improving farmed fish quality and safety** *Edited by Ø. Lie*
163 **Carbohydrate-active enzymes** *Edited by K.-H. Park*
164 **Chilled foods: a comprehensive guide Third edition** *Edited by M. Brown*
165 **Food for the ageing population** *Edited by M. M. Raats, C. P. G. M. de Groot and W. A Van Staveren*
166 **Improving the sensory and nutritional quality of fresh meat** *Edited by J. P. Kerry and D. A. Ledward*
167 **Shellfish safety and quality** *Edited by S. E. Shumway and G. E. Rodrick*
168 **Functional and speciality beverage technology** *Edited by P. Paquin*
169 **Functional foods: principles and technology** *M. Guo*
170 **Endocrine-disrupting chemicals in food** *Edited by I. Shaw*
171 **Meals in science and practice: interdisciplinary research and business applications** *Edited by H. L. Meiselman*
172 **Food constituents and oral health: current status and future prospects** *Edited by M. Wilson*
173 **Handbook of hydrocolloids Second edition** *Edited by G. O. Phillips and P. A. Williams*
174 **Food processing technology: principles and practice Third edition** *P. J. Fellows*
175 **Science and technology of enrobed and filled chocolate, confectionery and bakery products** *Edited by G. Talbot*
176 **Foodborne pathogens: hazards, risk analysis and control Second edition** *Edited by C. de W. Blackburn and P. J. McClure*
177 **Designing functional foods: measuring and controlling food structure breakdown and absorption** *Edited by D. J. McClements and E. A. Decker*
178 **New technologies in aquaculture: improving production efficiency, quality and environmental management** *Edited by G. Burnell and G. Allan*

179 **More baking problems solved** *S. P. Cauvain and L. S. Young*
180 **Soft drink and fruit juice problems solved** *P. Ashurst and R. Hargitt*
181 **Biofilms in the food and beverage industries** *Edited by P. M. Fratamico, B. A. Annous and N. W. Gunther*
182 **Dairy-derived ingredients: food and neutraceutical uses** *Edited by M. Corredig*
183 **Handbook of waste management and co-product recovery in food processing Volume 2** *Edited by K. W. Waldron*
184 **Innovations in food labelling** *Edited by J. Albert*
185 **Delivering performance in food supply chains** *Edited by C. Mena and G. Stevens*
186 **Chemical deterioration and physical instability of food and beverages** *Edited by L. H. Skibsted, J. Risbo and M. L. Andersen*
187 **Managing wine quality Volume 1: viticulture and wine quality** *Edited by A. G. Reynolds*
188 **Improving the safety and quality of milk Volume 1: milk production and processing** *Edited by M. Griffiths*
189 **Improving the safety and quality of milk Volume 2: improving quality in milk products** *Edited by M. Griffiths*
190 **Cereal grains: assessing and managing quality** *Edited by C. Wrigley and I. Batey*
191 **Sensory analysis for food and beverage quality control: a practical guide** *Edited by D. Kilcast*
192 **Managing wine quality Volume 2: oenology and wine quality** *Edited by A. G. Reynolds*
193 **Winemaking problems solved** *Edited by C. E. Butzke*
194 **Environmental assessment and management in the food industry** *Edited by U. Sonesson, J. Berlin and F. Ziegler*
195 **Consumer-driven innovation in food and personal care products** *Edited by S. R. Jaeger and H. MacFie*
196 **Tracing pathogens in the food chain** *Edited by S. Brul, P.M. Fratamico and T. A. McMeekin*
197 **Case studies in novel food processing technologies: innovations in processing, packaging, and predictive modelling** *Edited by C. J. Doona, K. Kustin and F. E. Feeherry*
198 **Freeze-drying of pharmaceutical and food products** *T.-C. Hua, B.-L. Liu and H. Zhang*
199 **Oxidation in foods and beverages and antioxidant applications Volume 1: understanding mechanisms of oxidation and antioxidant activity** *Edited by E. A. Decker, R. J. Elias and D. J. McClements*
200 **Oxidation in foods and beverages and antioxidant applications Volume 2: management in different industry sectors** *Edited by E. A. Decker, R. J. Elias and D. J. McClements*
201 **Protective cultures, antimicrobial metabolites and bacteriophages for food and beverage biopreservation** *Edited by C. Lacroix*
202 **Separation, extraction and concentration processes in the food, beverage and nutraceutical industries** *Edited by S. S. H. Rizvi*
203 **Determining mycotoxins and mycotoxigenic fungi in food and feed** *Edited by S. De Saeger*
204 **Developing children's food products** *Edited by D. Kilcast and F. Angus*
205 **Functional foods: concept to product Second edition** *Edited by M. Saarela*
206 **Postharvest biology and technology of tropical and subtropical fruits Volume 1: Fundamental issues** *Edited by E. M. Yahia*
207 **Postharvest biology and technology of tropical and subtropical fruits Volume 2: Açai to citrus** *Edited by E. M. Yahia*

208 **Postharvest biology and technology of tropical and subtropical fruits Volume 3:** Cocona to mango *Edited by E. M. Yahia*
209 **Postharvest biology and technology of tropical and subtropical fruits Volume 4:** Mangosteen to white sapote *Edited by E. M. Yahia*
210 **Food and beverage stability and shelf life** *Edited by D. Kilcast and P. Subramaniam*
211 **Processed Meats: improving safety, nutrition and quality** *Edited by J. P. Kerry and J. F. Kerry*
212 **Food chain integrity: a holistic approach to food traceability, safety, quality and authenticity** *Edited by J. Hoorfar, K. Jordan, F. Butler and R. Prugger*
213 **Improving the safety and quality of eggs and egg products Volume 1** *Edited by Y. Nys, M. Bain and F. Van Immerseel*
214 **Improving the safety and quality of eggs and egg products Volume 2** *Edited by F. Van Immerseel, Y. Nys and M. Bain*
215 **Animal feed contamination: effects on livestock and food safety** *Edited by J. Fink-Gremmels*
216 **Hygienic design of food factories** *Edited by J. Holah and H. L. M. Lelieveld*
217 **Manley's technology of biscuits, crackers and cookies Fourth edition** *Edited by D. Manley*
218 **Nanotechnology in the food, beverage and nutraceutical industries** *Edited by Q. Huang*
219 **Rice quality: A guide to rice properties and analysis** *K. R. Bhattacharya*
220 **Meat, poultry and seafood packaging** *Edited by J. P. Kerry*
221 **Reducing saturated fats in foods** *Edited by G. Talbot*
222 **Handbook of food proteins** *Edited by G. O. Phillips and P. A. Williams*
223 **Lifetime nutritional influences on cognition, behaviour and psychiatric illness** *Edited by D. Benton*
224 **Food machinery for the production of cereal foods, snack foods and confectionery** *L.-M. Cheng*
225 **Alcoholic beverages: sensory evaluation and consumer research** *Edited by J. Piggott*
226 **Extrusion problems solved: food, pet food and feed** *M. N. Riaz and G. J. Rokey*
227 **Handbook of herbs and spices Volume 1 Second edition** *Edited by K. V. Peter*
228 **Handbook of herbs and spices Volume 2 Second edition** *Edited by K. V. Peter*
229 **Bread making: improving quality Second edition** *Edited by S. P. Cauvain*
230 **Emerging food packaging technologies: principles and practice** *Edited by K. L. Yam and D. S. Lee*
231 **Infectious disease in aquaculture: prevention and control** *Edited by B. Austin*
232 **Diet, immunity and inflammation** *Edited by P. C. Calder and P. Yaqoob*
233 **Natural food additives, ingredients and flavourings** *Edited by D. Baines and R. Seal*
234 **Food decontamination: novel methods and applications** *Edited by A. Demirci and M. Ngadi*
235 **Chemical contaminants and residues in foods** *Edited by D. Schrenk*
236 **Robotics and automation in the food industry: current and future technologies** *Edited by D. G. Caldwell*
237 **Fibre-rich and wholegrain foods: improving quality** *Edited by J. A. Delcour and K. Poutanen*
238 **Computer vision technology in the food and beverage industries** *Edited by D.-W. Sun*
239 **Encapsulation technologies and delivery systems for food ingredients and nutraceuticals** *Edited by N. Garti and D. J. McClements*

Preface

Extrusion cooking has been used in the food and feed industries for many years to produce various extruded products. Among all of the thermal cooking processes utilized in the food and feed processing industries, extrusion is one of those that is most efficient and most commonly employed. The growing number of individuals associated with the extrusion process justifies a compilation of information regarding this topic. Major universities and colleges recognize a deficiency in educational material and are beginning to examine the options in order to offer an undergraduate curriculum that targets this segment of the food and feed industries.

The objective of this book is to answer the most frequently asked questions in the area of food and feed extrusion. It is based on questions that arise most commonly during the Food Protein Research and Development Center, Texas A&M University's short courses in the area of food and feed extrusion. The questions are divided into various sections so that relevant information can be accessed quickly. These sections include: components of a food or feed extrusion system, selecting the right type of extruder, impact of protein, starch, fat and fiber on extruded foods and feeds preconditioners, die and knife assemblies, drying and practical considerations, rules of thumb and solutions to the most common problems in food and feed extrusion. Discussions about specific extruders in this book does not constitute endorsement or preference of products or services by The Texas A&M University System or its divisions.

The authors of this book have extensive experience (more than 57 years combined) in the area of extrusion cooking of food and feed. This book can be used by anyone who is interested in learning about extrusion as well as by someone who is already working in the extrusion field. More than three hundred questions and answers are provided which can solve many issues that extruder operators can face in their daily activities and because of their responsibilities. Much of the subject content may be familiar to experienced, well-seasoned production personnel but can provide a different perspective on many common process challenges. For individuals just entering the field of extrusion, the 'question/answer' approach is intended to be an invaluable resource, encompassing the most frequently asked questions in a user-friendly format. We hope this book will bring valuable information to and answer most questions of those who are involved in the area of food and feed extrusion.

Special thanks are due to Mr Tom Barber from Buhler-Aeroglide who provided answers to the questions related to drying of extruded products. We would also like to thank Ms Sarah Whitworth, Senior Commissioning Editor (Food science, technology and nutrition) at Woodhead publishing for providing us guidance and suggestions related to this book.

Mian N. Riaz
Galen J. Rokey

Sources of further information

Various information sources on extrusion technology are available, including books, chapters in food and feed books and magazines which contain articles on extrusion. Various extruder manufacturers also have useful articles and information on their website.

Books

CHANG, Y. K. AND WANG, S. S. (eds) (1999). *Advances in Extrusion Technology: Aquaculture/ Animal Feeds and Foods*, Technomic Publishing, Lancaster, Pennsylvania.

FAST, R. B. AND CALDWELL, E. F. (eds) (2000). *Breakfast Cereals and How They are Made*, 2nd edn. American Association of Cereal Chemists, St. Paul, Minnesota.

FRAME, N. D. (ed.) (1994). *The Technology of Extrusion Cooking*, Blackie Academic & Professional, New York.

GUY, R. (ed.) (2001). *Extrusion Cooking Technologies and Application*, Woodhead Publishing Limited, Cambridge, UK.

HARPER, J. M. (1981). *Extrusion of Foods. Vol. 1 & 2*, CRC Press, Baco Raton, Florida.

HAYAKAWA, I. (ed.) (1992). *Food Processing by Ultra High Pressure Twin-screw Extrusion., I*, Technomic Publishing, Lancaster, Pennsylvania.

JANSSEN, L. P. B. M. (ed.) (1978). *Twin-screw Extrusion*. Elsevier Applied Science, New York.

JOWITT, R. (1984). *Extrusion Cooking Technology*, Elsevier Applied Science, New York.

KOKINI, J. L., HO, C. AND KARWE, M. V. (eds) (1992). *Food Extrusion Science and Technology*. Marcel Dekker, New York.

MERCIER, C. AND CANTARELLI, C. (eds) (1986). *Pasta and Extrusion Cooked Foods*, Elsevier Applied Science, New York.

MERCIER, C., LINKO, P. AND HARPER, J. M. (eds) (1989). *Extrusion Cooking*, American Association of Cereal Chemist, St. Paul, Minnesota.

O'CONNOR, C. (ed.) (1987). *Extrusion Technology For the Food Industry*, Elsevier Applied Science, New York.

RIAZ, M. N. (ed.) (2000). *Extruders in Food Applications*, Technomic Publishing, PA (bought by CRC Press).

RIAZ, M. N. (ed.) (2007). *Extruders and Expanders in Pet Food, Aquatic and Livestock Feeds*, AgriMedia, Clenze, Germany.

SCHOLFIELD, E. K. (ed.) (2005). *Feed Manufacturing Technology V*, American Feed Industry Association, Virginia, USA.

WILSON, D. AND TRIBELHORN, R. E. (eds) (1979). *Low-cost Extrusion Cookers, Workshop Proceedings*, United States Department of Agriculture. Office of International Cooperation and Development, Washington DC.

WOODROOFE, J. M. (1993). *Dry Extrusion Manual*, Rural Pacific Pty, Australia.

ZEUTHEN, P., CHEFTEL, J. C., ERIKSSON, C., JUL, M., LENIGER, H., LINKO, P., VARELA, G. AND VOS, G. (eds) (1984). *Thermal Processing and Quality of Foods*, Elsevier Applied Science, New York

Focus on Extrusion (1999). *Proceedings of the Pet Food Industry*, Chicago, Illinois, 24–25 March, 1999. Watt Publishing, Mt Morris, Il.

Book chapters

RIAZ, M. N. (2005). 'Extrusion processing of oilseed meals for food and feed production', in *Bailey's Industrial Oil and Fat Products. Vol 5, Edible Oil and Fat Products: Processing Technologies*, 6th edn, John Wiley & Sons, Cutten, CA., and Shahidi F. (ed.), Chapter 13, pp 533–72.

RIAZ, M. N. (2006). 'Extruded snacks', in *Handbook of Food Science, Technology, and Engineering*, Hui, Y. H. (ed.), Taylor and Francis, CRC Press, Boca Raton, FL, Vol. 4, pp 168-1–168-8.

RIAZ, M. N. (2008). 'Extrusion of aquafeeds formulated with alternative protein sources', in *Alternative Protein Sources in Aquaculture Diets*, Lim, C., Webster, C. D. and Lee, C.-S. (eds), The Haworth Press, Taylor & Francis Group, New York, NY, 523–46.

RIAZ, M. N. AND WILLIAMS, M. (2003). 'Extrusion basics', p 347–360. in *Pet Food Technology*, Kvamme, J. and Phillips, T. (eds), Watt Publishing Co, Mt. Morris, IL, 347–60.

Magazines and journals

All About Feed, Elsevier International, The Netherlands.
Cereal Food Worlds, American Association of Cereal Chemists, St. Paul, Minnesota.
Feed International, Watts Publications, Watt Publishing Co, Rockford, Illinois.
Feed Management, Watts Publications, Watt Publishing Co, Rockford, Illinois.
Food Processing, Putnam Publications, Itasca, Illinois.
Food Technology, Institute of Food Technologists, Chicago, Illinois.
International Aquafeed Directory, Turret RAI, United Kingdom.
Petfood Industry, Watts Publications, Watt Publishing Co, Rockford, Illinois.
Prepared Foods, Cahners Food Group Publications, Oak Brook, Illinois.

An introduction to food and feed extrusion and associated terminology

1 What is extrusion cooking?

Extrusion cooking is a process in which a food or feed material is forced to flow, under elevated temperature, pressure and shear, through a die.

2 What is cold extrusion?

Cold extrusion refers to cooking of food or feed ingredients at temperatures below 40 °C. The maximum process temperature could be higher in some applications but always stays below 100 °C. If temperatures remain below this level, no steam is formed.

3 What is dry extrusion?

Dry extrusion is when cooking is achieved through friction and at moisture levels that result in final extrudate moistures below 12%.

4 What are the functions of a typical extruder?

Depending upon the type of extruder, different functions can be performed. These include agglomeration of ingredients, expansion, gelatinization, homogenization, dehydration, mixing, pasteurization, protein denaturation, shearing, texture alteration and product shaping.

5 What are the advantages of extrusion compared to other thermal processes?

Extrusion cooking offers several advantages: a variety of products can be made by changing the ingredients and operating conditions, multiple shapes can be created, a variety of textures and colors can be produced, cooking is energy efficient, processing costs are lower, a variety of raw materials from all sources (animal and plant) can be processed, and full automation is a possibility.

6 What is the quality of the products produced by extrusion?

Since extrusion is a high temperature short time (HTST) heating process, it minimizes degradation of food nutrients while improving the digestibility of proteins (by denaturation) and starches (by gelatinization). Extrusion cooking at high temperature also destroys anti-nutritional compounds, that is trypsin inhibitors, and undesirable enzymes, such as lipases, lipoxidases and micro-organisms.

7 What is the waste stream from extrusion processing?

Extrusion produces few or no waste streams. This is a very important advantage for the food and feed industries as new environmental regulations are stringent and costly.

8 How easy is it to scale up the extrusion process?

Data obtained from the pilot plant extruders can be used to scale up the extrusion system for commercial production.

9 Can I use extruders as continuous reactors?

Extruders are being used as continuous reactors in several countries for deactivation of aflatoxin in peanut meals and destruction of allergens and toxic compounds in castor seed meal and other oilseed crops.

10 Do extruders need to be shut down every day?

Typically, production extruders operate 24 hours a day, and 'shutdowns' for cleanup do not occur except for required sanitation or maintenance.

11 What is feedstock in extrusion?

Feedstock is the material, mixture, or formulation to be processed in extruder.

12 What is a pellet?

A pellet is a discrete particle which is shaped and cut by the extruder. It is called a pellet regardless of its shape.

13 What is the recipe or formulation?

The recipe is the ingredients or mixture to be processed in an extruder to make food, aquatic and livestock feed.

14 What is the compression ratio?

The compression ratio (CR) is the volume of the full flight of the screw at the feed opening divided by volume of the last full flight before discharge. Typical CR ranges are from 1:1 to 5:1.

15 What is *L/D* (length to diameter) ratio?

This is also called 'L' over 'D'. It is the distance from the internal rear edge to the discharge end of the barrel, divided by the diameter of the bore. Food extruder *L/D* ratios range from 1:1 to 25:1 (see Fig. 1).

16 What is screw geometry?

The term screw geometry refers to the screw profile within the extruder. It is a combination of flight lead or pitch, flight angle, flight depth, and flight diameter. The geometry can easily be varied on extruder designs where the screw elements are segmented and driven by a central shaft. Other designs have the screw geometry (screw profile) machined into the shaft and this reduces flexibility in altering the screw geometry as the whole shaft must be removed and replaced with another, one-piece shaft with a different screw geometry machined into the shaft.

17 What does pitch mean?

This is the angle of the flight relative to the axis of the root of the screw (see Fig. 2).

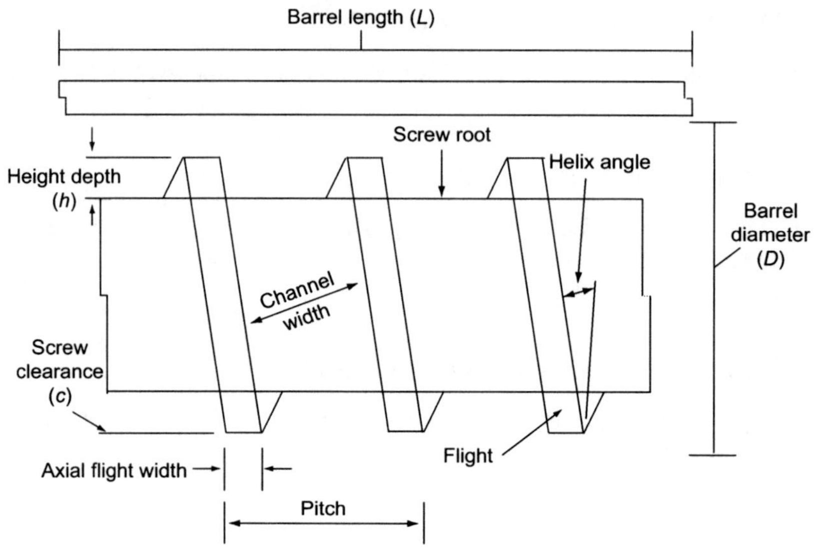

Fig. 1 Barrel and other components.

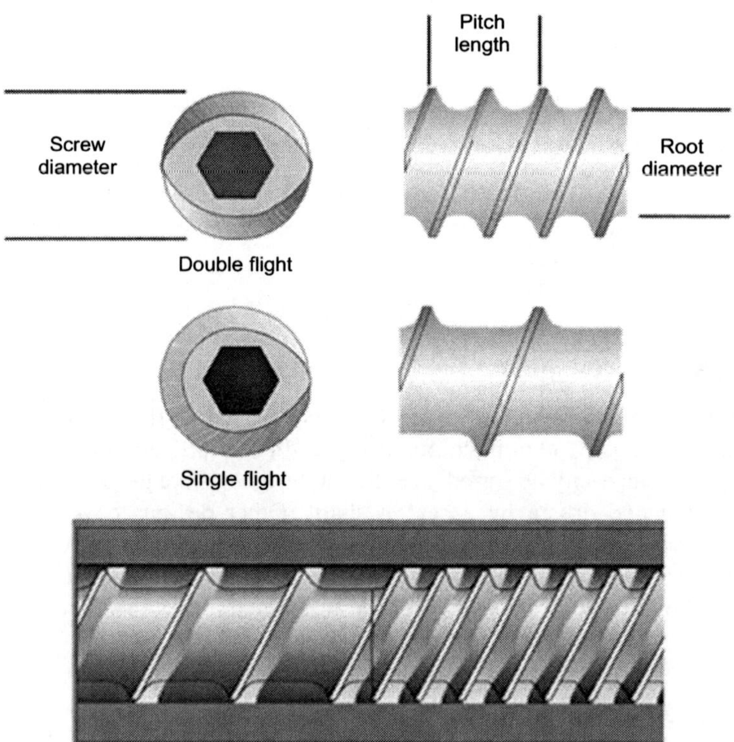

Fig. 2 Screw terminology.

18 What is the flight depth?

The flight depth is determined by taking the outside diameter or bore of the screw, subtracting the root diameter of the screw and dividing by two. In other words, it the distance from the tip of the screw flighting to the root diameter of the screw.

19 What is the flight angle?

The flight angle is the angle described by the plane of the leading edge of the screw flighting with respect to a plane that is perpendicular to the centerline of the extruder shaft.

20 What does 'venting concept' mean?

This is used when we need to make a dense food or sinking aquaculture feed. Towards the end, one of the barrel sections will have an opening to release the pressure and steam to densify the food, aquatic and livestock feed (see Fig. 3).

21 What does shear mean?

This is the working mixing action that homogenizes and heats the raw ingredients. It is often described as the force applied to a material or an object over a given distance.

Fig. 3 Extruder vent.

22 What are long-cut and short-cut products?

Long-cut and short-cut are terms describing the length of the extrudate in relationship to its diameter. Generally long-cut products are referred to as 'long goods' and are at least twice their diameter in length. Usually long goods are extruded as continuous ropes which are conveyed away from the die in belt conveyors. As the product travels along the conveyor it is cooled and the structure becomes more rigid. This 'setting' of the product allows the product to be cut into long lengths by a mechanical cutter (usually a guillotine device) to lengths that are at least twice the diameter. Short-cut products or short goods are usually cut immediately at the die face by a rotary knife. The shorter length of this product means that it does not become distorted by gravity. If a long-goods product were face cut at the die (as happens in the manufacture of short goods), the product would become curved or distorted in shape along its length due to gravity as it protrudes from the die orifice before being cut to length by the knife.

23 What does collet mean?

A collet is an extrudate described as being a small, finger-like product. In oilseed extrusion it is the coarse pieces made when extruding oilseeds to enhance their solvent extraction characteristics (see Fig. 4).

Fig. 4 Collets.

24 What is kibble?

Dog and cat food are sometimes also known as kibble.

25 What is starch gelatinization?

Gelatinization is the cooking of starch. When starch becomes gelatinized, its crystalline areas melt, the granules take on water and swell, viscosity increases and the granules become fragile and susceptible to digestion by enzymes. Gelatinization requires the presence of heat, moisture, and time.

26 What is the gelatinization temperature of starch?

The gelatinization temperature varies for different starches. For example, barley starch gelatinization takes place at 56 °C but the gelatinization temperature for sorghum starch it is 73 °C. Table 1 presents some other starches and their gelatinization temperatures.

27 What is dextrinization?

If starch is subjected to dry heat, it breaks down to form pyrodextrins in a process known as dextrinization. Pyrodextrins are brown in color. This process is partially responsible for the browning of toasted bread.

28 What is the Maillard reaction?

The Maillard reaction is a chemical reaction between an amino acid and a reducing

Table 1 Gelatinization temperature of different starches

Starch type	Gelatinization temperature (°C)
Triticale	59
Wheat	61
Rye	64
Oats	56
Corn	67
Waxy corn	68
Rice (short)	68
Rice (medium)	68
Rice (long)	71
Potato	61
Tapioca	65

sugar, usually requiring heat. It is vitally important in the preparation or presentation of many types of food.

29 What is meant by retention time?

The retention time is the time the raw material stays in the extruder barrel. Typically it ranges from 15 to 30 seconds.

30 What is surging?

Surging means extruder instability. The motor load of the extruder is going up and down and is not stable. This is usually due to non-uniform delivery of a dry recipe or any of the other mass flow inputs such as water, steam or liquid additives. There could be several other reasons for surging, such as the barrel temperature being too high, the open area in the die being too large, or steam injection to the extruder being too high.

31 What are fines?

During extrusion and drying some material comes off the pellets or kibble. This material is called fines.

32 What is meant by grind?

Grind means the particle size of the raw material after grinding.

33 What is meant by HTST?

HTST is the most common form of pasteurization in the dairy industry. HTST stands for high temperature/short time, but the process is referred to most frequently by its acronym. Basically HTST means that the milk is heated to a required minimum temperature of 72 °C for 15 seconds. Extrusion cooking is a HTST process too. In extrusion cooking the temperature could be as high as 200 °C and could be held for 5–10 seconds.

34 What is a PLC?

A PLC is a programmable logic controller.

35 What is the glass transition?

Glass transition is what occurs when enough of a material has softened to allow for particle compaction.

36 What is the melt phase transition?

Melt phase transition is what occurs when enough of a material has melted to allow for melt flow.

Components of a food or feed extrusion system

37 What is a feeder?

This device ensures uniform delivery of the feed and pet food ingredients to the preconditioner or extruder. Feeders are either volumetric or variable-speed augers.

38 What is a volumetric feeder?

'Volumetric' means measurement by volume. In a volumetric feeder, the auger is the means used to measure a certain volume of product. This volume of product is discharged with each revolution of the auger. In this type of feeder the operator sets the feeder speed to control the feed rate. However, the actual feed rate will be affected by changes in bulk density of the raw material. The design of a volumetric feeder is very simple and it is a low cost piece of equipment. The bin fill rate normally required is five times the extrusion rate.

39 What is a gravimetric feeder?

'Gravimetric' means measurement by weight. In this type of feeder the operator sets the feed rate but the system is under automatic control, operating on the basis of the loss of a specific amount of weight of product in a certain time period. These types of feeder are not affected by changes in the bulk density of the raw material, so are more precise than volumetric feeders. Their design is more complex and because they have more of a control system they are more expensive. Gravimetric feeders are also known as variable-speed augers.

Fig. 1 Feeding bin.

40 What is a hopper or bin?

A hopper or bin holds the feed or pet food ingredients above the feeder. It is usually made from stainless steel with a glass window so that the level of the ingredients is visible (see Fig. 1).

41 What is a preconditioner?

This part is inbetween the feeder and the extruder barrel. Moisture and steam can be added in the preconditioner to partially cook and mix the raw material before it goes into the extruder barrel. Not all extruders have a preconditioner (see Fig. 2).

42 What is a diverter spout?

The term diverter spout generally refers to a device located between the discharge of the preconditioner and the inlet of the extruder. The diverter spout provides a way for the extruder operator to divert product coming out of the preconditioner before it goes into the extruder. It can be used to check that a recipe is properly preconditioned before the flow stream is channeled into the extruder. It also provides a method of sampling product being discharged from the preconditioner.

Fig. 2 Preconditioner.

43 What is an extruder barrel?

This key part of an extruder is between the preconditioner and the knife assembly. An extruder barrel contains shafts (single or twin) and rotating elements (screws and shear locks) for cooking and processing the food, feed or pet food (see Fig. 3).

44 What does the phrase 'barrel section' mean?

A section of the barrel, also called a segment, is a portion of the barrel that can be removed and swapped for another one if necessary. It may contain its own cooling/heating sleeve. The segments may contain grooving or spiraling and often are the same length as the worm screws. Barrel sections make the extrusion process more flexible, as different types of section can be assembled together to enhance the effects of the worm segments they contain (see Fig. 4).

Fig. 3 Extruder barrel.

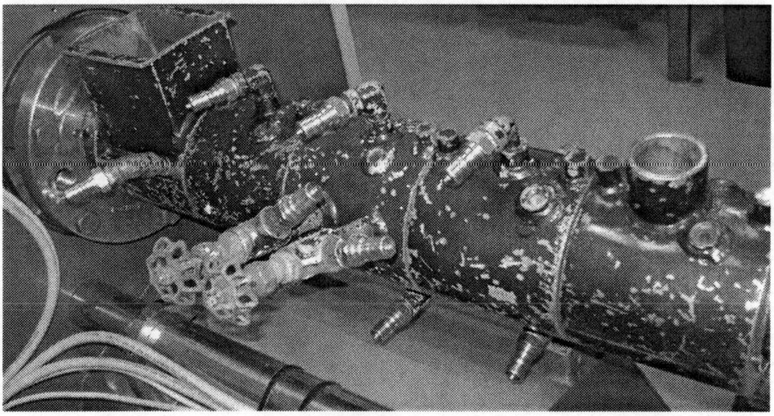

Fig. 4 Barrel sections.

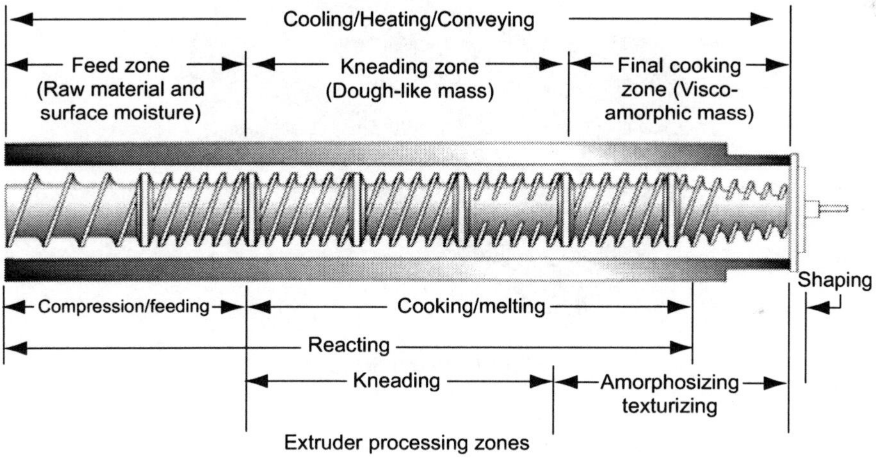

Fig. 5 Feeding zone.

45 What is the feeding section of the barrel?

This section usually contains deep flight screws to carry the raw material immediately to the next section. The main function of the feeding section is to ensure that raw material moves quickly to the barrel. Otherwise, the barrel will plug up and the extruder will have to be shut down (see Fig. 5).

46 What is the compression section of the barrel?

The main function of the compression section is to plasticize the raw material into a dough-like state. This section is often located between the feeding section and

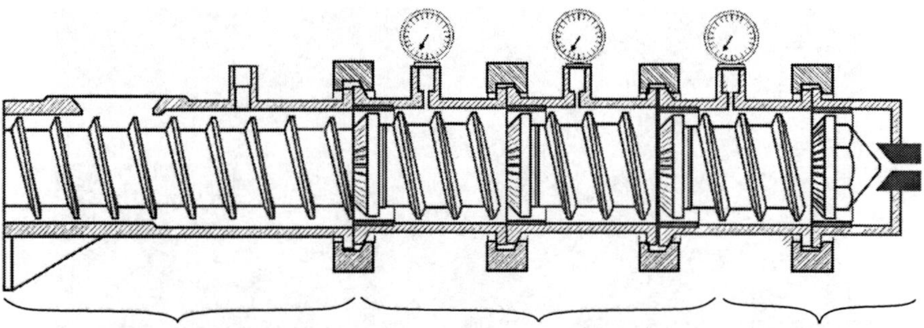

Fig. 6 Metering zone (courtesy of Insta-Pro).

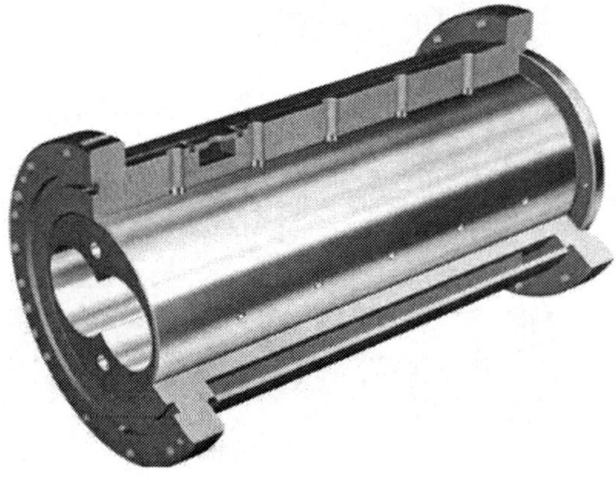

Fig. 7 Barrel liner.

metering sections. There can be a gradual decrease in the screw flight depth in the direction of die in this section.

47 What is the metering section of the barrel?

This section is nearest the discharge end of the extruder. The main function of the metering section is to increase the shear rate and cooking of the raw material and to 'meter' or uniformly feed the product into the die area. This section may have a very shallow flight screw (see Fig. 6).

48 What is a barrel liner?

The removable sleeve within the barrel section is called the liner. Usually the liner helps with transport of product and can be replaced when worn (see Fig. 7).

Fig. 8 Jacket on the barrel to cool the product.

49 What are rifling grooves?

Rifling refers to the helical grooves in the liner or inside of the extruder barrel. The rifling or spiral groove prevents product from slipping in a single screw extruder design and promotes conveying.

50 What does the term 'jacket' mean?

Jackets are hollow cavities or sleeves outside or around the barrel in which cooling or heating media such as water, oil or steam can be run to cool or heat the barrel surface (see Fig. 8). They may also be referred to as cooling/heating jackets. In some locations, direct electrical heating of the barrel may be desirable.

51 What is a screw?

The screw accepts the feed ingredients at the feed port, conveys, works and then forces them through the die (see Fig. 9). There are several different types of screws, including single flight screws, double flight screws, cut flight screws, shallow flight screws, deep flight screws, and feeding screws. Screw design differs from manufacturer to manufacturer and from one type of extruder to another.

52 What are screw crests and troughs?

The term screw crest refers to the tip of the helical flighting or the largest outside

Fig. 9 Screws (courtesy of Insta-Pro).

diameter of the screw. Screw troughs then are the voids between screw flightings and extend from the tip of the screw flighting down into the root diameter of the screw.

53 What is the root?

This is the continuous central shaft of the screw. The screw is usually a cylindrical or conical shape, around which the flight is wound (see Figs 10 and 11).

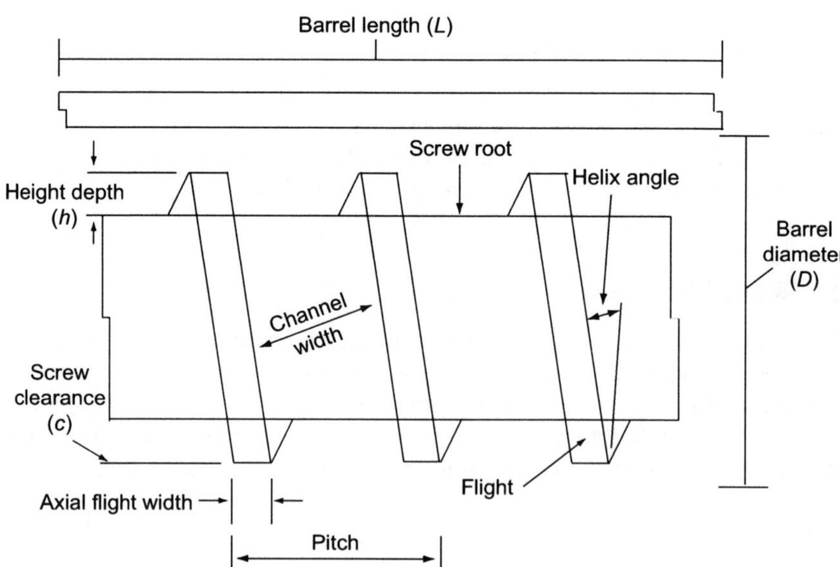

Fig. 10 Barrel and other components.

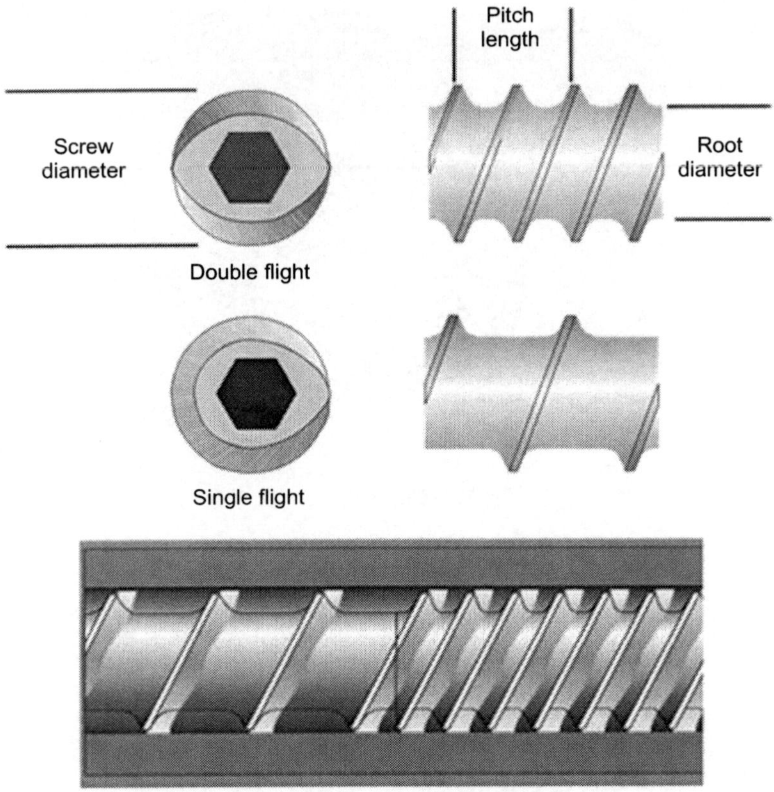

Fig. 11 Screw terminology.

54 What is a screw element?

A screw element is a detachable segment of a segmented screw. A segmented screw is a screw that is modular, that is it consists of various screw elements.

55 What is a hollow-core screw?

A hollow-core screw is a solid screw or shaft that has been drilled to open up its core for the circulation of heating or cooling liquids, thus providing extra heat transfer surface area.

56 What is a screw flight?

A screw flight (or simply flight) is the helical conveying surface of the screw, which pushes the product forward (see Figs 10 and 11).

Fig. 12 Cut flight screws.

57 What is a continuous flight?

Continuous flighting in an extruder screw means there are no interruptions in the flighting. One can follow the flighting from the beginning to the end of the screw as it forms a continuous helix around the root diameter of the shaft.

58 What is a cut flight screw?

A cut flight screw is also called an interrupted flight screw. In a screw of this type, a section of the flight is missing. The fact that it is missing increases the shearing action, increasing cooking of the raw ingredients (see Fig. 12).

59 What is an interrupted flight screw?

Interrupted flight screw is another word for cut flight screw. Usually, studs (bolts) are inserted through the extruder barrel wall into the empty flight section to induce more shear in the product. Steam may also be injected into the product through valves placed in the stud holes.

60 What is the screw bore?

The screw bore is the outside diameter of the screw flighting.

61 What is a worm?

A worm is a hollow-core segmented screw element that slips over the shaft in a

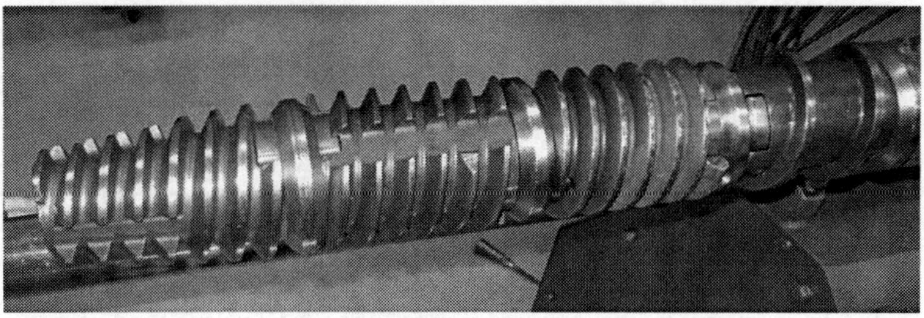

Fig. 13 Worm.

Fig. 14 Shear rings.

modular screw. Screws with various profiles and actions can be assembled by selecting appropriate worm sections (see Fig. 13).

62 What is a shear ring?

A shear ring is sometimes also called shear lock or steam lock or ring dam. This is a round device that may lock together individual screws in the barrel and serves as a restrictive device to control shear and to prevent steam and product from flowing back into the inlet area (see Fig. 14).

Fig. 15 Extruder vent.

63 What are shearing bolts and what are they for?

Shearing bolts are threaded into and through the side of the extruder barrel wall and are threaded to allow adjustment of their degree of penetration into the extruder. The screw flighting must be interrupted to prevent interference with the shearing bolts. The distance the shearing bolts extend into the extruder and the more bolts that are engaged determines the amount of 'shear' imparted to the product.

64 What is a vent in the context of extrusion?

A vent is an opening in the extruder barrel before the die plate which allows pressure and steam to be removed from the product (see Fig. 15).

65 What is a die?

A die is a plate at the end of the extruder that contains small openings to shape the products (see Fig. 16).

66 What is a die insert?

Individual die openings can exist in plugs called inserts which slip into receptacles in a die plate (see Fig. 17).

Fig. 16 Die.

Fig. 17 Die inserts.

67 What is die land length?

This is a ratio between the length and diameter of the die orifice through which the product passes. Longer die land length provides more back pressure and may have an impact on product density (see Fig. 18).

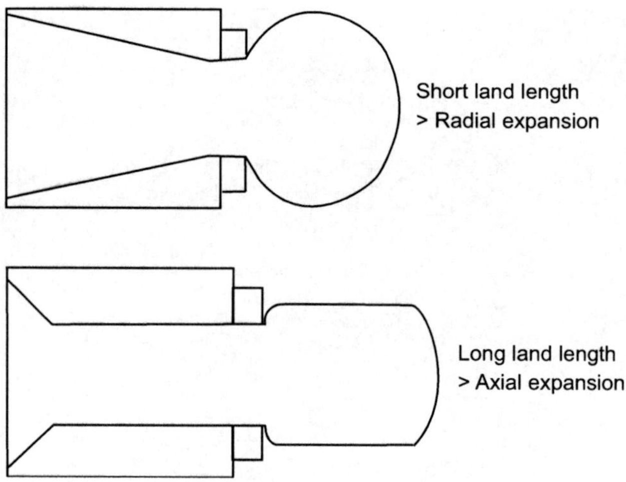

Fig. 18 Land lengths of dies.

68 What is the die orifice?

Die orifice is the term used to denote the openings in the plate at the end of the extruder barrel. The extrudate flows through the die orifice into the environment outside the extruder barrel. The shape of the orifice largely determines the shape of the extrudate.

Fig. 19 Die plate.

69 What is a die plate?

A die plate contains several holes which can receive individual die inserts containing the actual die opening. Die holes may be drilled directly into the plate or the plate may be machined to hold die inserts that have complex designs for shaping the product and may be constructed from hard-wearing material (see Fig. 19).

70 What is a choke plate?

A choke plate is a plate installed between the end of the screw and the final die plate. The choke plate can contain any number of holes of various sizes depending on the amount of restriction desired. Fewer and smaller openings in the choke plate increase restriction of material flow within the extruder. The increased restriction in turn will increase barrel fill, mixing and cook.

71 What is a single-face die plate?

A single face die plate is a flat plate attached to the discharge of an extruder barrel. This single plate contains openings through which the extrudate flows from the extruder into the atmosphere. There are circumstances where multiple die plates are utilized, which are separated by predetermined spaces and also contain orifices through which the product flows. The use of multiple plates increases the restriction to flow of the product, thus serving to increase cook and barrel fill. Most commonly, though, extrusion applications utilize a single die plate as this is sufficient to restrict the flow of the extrudate.

72 What is a breaker plate?

A breaker plate is usually located between the die and the end of the extruder and

Fig. 20 Breaker plate.

serves as a restrictive device to providing uniform product flow to the die inlet (see Fig. 20).

73 What is the cutter?

The cutter, also called a knife cutter, is the assembly which cuts the extrudate into pieces of the desired length (see Figs 21 and 22).

Fig. 21 Cutter (courtesy of Insta-Pro).

Fig. 22 Knife cutter.

Selecting the right type of extruder: single screw and twin screw extruders for food and feed production

74 What are the different types of extruders that are available?

Extruders can be classified in different ways; however, they are most commonly classified based on screw design as follows:

- Single screw extruders (Fig. 1).
- Twin screw extruders (Fig. 2).

Extruders can also be classified based on the source of heat and level of moisture in the material during extrusion:

- Dry extruders.
- Wet extruders.

75 What is a single screw extruder?

A single screw extruder is an extruder that has one shaft in the barrel. The single screw extruder is the most widely applied extrusion device in the food industry. The single screw extruder also produces more tons of extruded products than any other extrusion processing method (see Fig. 1).

76 What is a wet extruder?

In wet extrusion, steam is injected and the product is usually dried after extrusion.

Fig. 1 Cutaway of a single screw extruder.

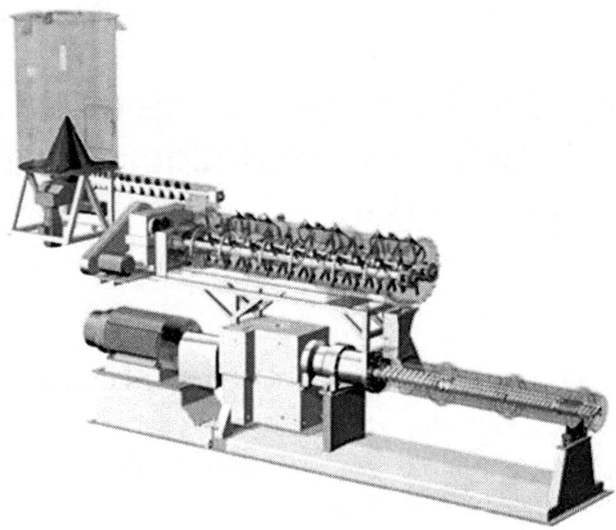

Fig. 2 Twin screw extrusion system.

77 What is a segmented screw/barrel single-screw 'wet' extruder?

A segmented screw/barrel single-screw extruder is the most widely used extruder design for extrusion cooking in the food, pet food and feed processing industries. As mentioned previously, 'wet' means that steam and water can be injected into the barrel during processing. Typically, the barrels of these machines are also equipped with heating and cooling jackets.

78 What does a typical single screw consist of?

A typical single screw extruder consists of a live bin, feeding screw, preconditioning cylinder, extruder barrel, die and knife (see Fig. 3)

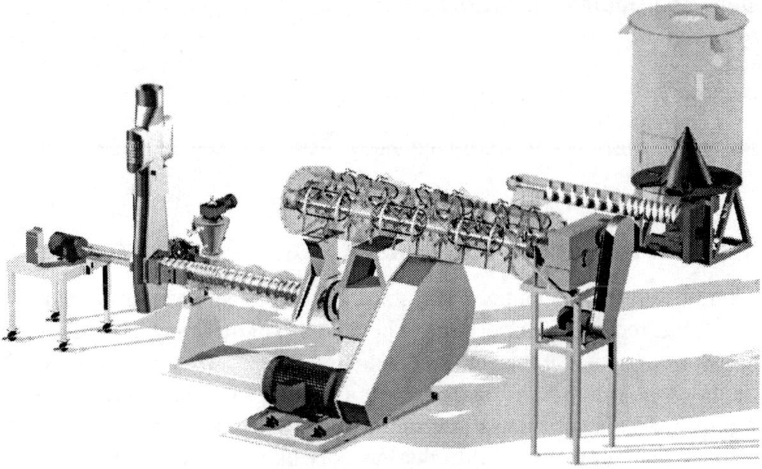

Fig. 3 A typical extruder.

79 How much internal fat can be present in a recipe to be processed by a single screw extruder?

Recipes containing more than 12% internal fat may cause slippage inside the barrel, resulting in less shear, a lower pressure and less cooking of the product.

80 What is the range of particle sizes that can be used in a single screw extruder?

Ingredient particle size distribution (sieve size) may be limiting. Very fine powders may not feed in this type of extruder well unless preconditioned and very coarse material will not cook properly.

81 What are the typical applications of single screw extruders?

A typical single screw extruder can process a variety of food and feed products. The following is a brief list of the products that are commonly made with single screw extruders:

- Direct expanded corn snacks.
- Texturized vegetable protein.
- Ready-to-eat-breakfast cereal.
- Full fat soy.
- Pet food.

- Floating and sinking aquatic feed.
- Baby foods.
- Breading.
- Precooked or thermally modified starches, flours and grain.

In addition rice bran can be stabilized using a single screw extruder.

82 What are the pros and cons of single screw extruders?

Pros: Single screw segmented 'wet' extruders are easy to operate and minimal training is required for the operators. The cost of single screw machines is about half that of twin-screw extruders and maintenance costs are also lower. Fewer complications exist in assembling screw configurations compared to those found in twin screw extruders because intermeshing between two screws is not required. **Cons:** Since the single screw extruder has only one shaft, it will not self-clean as completely at the end of the operation. However, if formulas of pet food or feed products are nearly similar, a prepared second swing-away die plate assembly can be mounted at the discharge end. Usually, the extruder can be stopped, the die plate loosened, swung out and replaced by the other in a very short downtime. If not allowed to cool, the extruder can be started up immediately. The opportunity may exist to add up to 10% of the start-up product as rework in succeeding runs. Also, increased water or oil and ground corn or cracked soybeans may be added to help clear the cooling extruder after shutdown by flushing the system with these materials. Final product diameter sizes can also be a factor, as fish feeds smaller than 1.5 mm may be more difficult to produce on a single screw extruder.

83 What are the different types of single screw extruder that are available?

Not all extruders are used for cooking and/or texturizing. Single screw extruders come in a number of sizes and shapes, and their screw, barrel and die configurations can usually be varied to suit a particular product's specification. There are several different types of single screw extruders available on the market. Examples are cold forming extruders, high pressure forming extruders, low shear cooking extruders, collect extruders and high shear cooking extruders.

84 What are cold forming extruders?

Cold forming extruders are low shear machines with smooth barrels, deep-flights and low screw speeds, originally used to work moistened semolina flour and press it through a die with little cooking during the pasta-making process. Similar extruders are used as continuous mixer-formers for the manufacture of pastry dough, cookies, processed meats and certain candies.

85 What are high-pressure forming extruders?

High-pressure forming extruders are low shear machines with grooved barrels and compressing screws, typically used to extrude pregelatinized cereal and other dough through dies to make pellets for subsequent drying and puffing or frying. Product temperatures are kept low to prevent unwanted puffing at the die. Various cereals and fried snack foods are made with these machines.

86 What is a low-shear cooking extruder?

Low-shear cooking extruders are moderate shear machines with high compression screws and grooved barrels to enhance mixing. Heat can be applied to the barrel or screw to 'cook' the product (i.e. pasteurize it, inactivate enzymes, denature proteins, gelatinize starch), while puffing at the die is generally avoided. Soft-moist foods and meat-like snacks, such as simulated jerky, can be made with these machines. The ingredients are often premixed to a dough-like consistency using other equipment.

87 What are collet extruders?

Collet extruders are high shear machines with grooved barrels and screws with multiple shallow flights that are used to make puffed snacks from defatted corn grits. The temperature of the relatively dry (12% moisture) ingredients is raised rapidly to over 175 °C and the starch is dextrinized and partially gelatinized. The resulting mass loses moisture and puffs immediately upon exit through a die to form a crisp, expanded curl or collet. This type of machine initially was characterized by an extremely short screw (length:diameter (*L*:*D*) = 3:1), but longer *L*/*D* (10:1) machines that rely heavily on friction-induced heat to produce collets have been developed. A 'collet-type' short *L*/*D* extruder is imported into the USA and offered domestically for processing animal feeds.

88 What are high-shear cooking extruders?

High-shear cooking extruders are high shear machines which can use screws with different flight depths and/or screw pitch, that have the ability to achieve high compression ratios, high temperatures and various degrees of puffing. Long barrel (length:diameter = 13–25:1) extruders sometimes adapted from the plastics industry were used initially, but many design modifications have been introduced for processing foods. A large variety of screw and internal barrel designs, and heating and cooling options exist. Some machines are equipped with conditioning chambers to premoisten and preheat the feedstock material.

89 What is a twin screw extruder?

The term 'twin screw' applies to extruders with two screws of equal length placed inside the same barrel. They are usually more expensive than single screw extruders. Twin-screw extruders are more complicated than single screw extruders, but at the same time provide much more flexibility and better control (see Fig. 2).

90 What are the different types of twin screw extruders that are available?

Twin screw extruders include a variety of machines with widely different processing and mechanical characteristics and capabilities. These extruders can be categorized according to the position of the screws in relation to one another and to the direction of screw rotation. In these types of extruder the screws either rotate in opposing directions (counter-rotating) or in the same direction (co-rotating). Regarding screw position, extruders can have:

- intermeshing screws in which the flight of one screw engages or penetrates the channels of the other screw, or
- non-intermeshing screws in which the screws do not engage each other's threads, allowing one screw to turn without interfering with the other. Non-intermeshing screw extruders function like single-screw extruders but potentially have a higher capacity.

Extruder screws can be either co-rotating, that is turn in the same direction with their screw crests matching up with opposing troughs, or counter-rotating, that is turn in opposite directions. Therefore four types of twin-screw extruders are possible:

1. non-intermeshed, co-rotating;
2. non-intermeshed, counter-rotating;
3. intermeshed, co-rotating;
4. intermeshed, counter-rotating.

91 What is the most common design used for twin screw extrusion?

The most common design used for twin screw extruders is intermeshed, co-rotating.

92 Are there any benefits of using twin screw versus a single screw extruder?

Twin screw extrusion offers several advantages over single screw extrusion processing. These advantages are:

- Twin screw extruders can handle viscous, oily, sticky or very wet material and some other products which will slip in a single screw extruder.
- A twin screw extruder has a positive pumping action and pulsation is reduced at the die. This means that very uniform product lengths and intricate shapes can be produced.
- Twin screw extruders may be less sensitive to wear than single screw extruders.
- Twin screw extruders can utilize a wide range of particle sizes (from fine powder to grains), whereas single screws are somewhat more limited to a specific range of particle sizes.
- Twin screw extruders are self-wiping, so cleanup is very easy. Since the twin-screw extruder contains two shafts, the inside of the barrel is swept clean and the two screws wipe each other. At the end of the operation, steam and water can usually clean the extruder barrel from the inside.
- Twin screw extrusion offers ease of process scale-up from pilot plant to large-scale production.
- If the end of the barrel of a twin screw extruder splits the output into two channels, products with pieces of two different colors or in a variegated form can be made by injecting color solutions into the channels just before the die.
- Three-dimensional or very delicate food products can be processed in a twin screw extruder and they are more forgiving to inexperienced operators.

93 What is the maximum fat level used in a twin screw extruder?

Usually, a twin screw extruder can handle products with greater than 18% internal fat compared to 12% maximum for single screw machines. The maximum fat level that can be used in a twin screw extruder is 22%.

94 What is the minimum size of the pellet that can be produced with a twin screw extruder?

A twin screw extruder can produce pellets of less than 1 mm diameter for food or feed applications; this latter capability is important in feeding fish fingerlings.

95 What is the range of particle sizes of raw material that can be used in a twin screw extruder?

Twin screw extruders can handle a wide range of the grind sizes of recipe ingredients. Very finely ground ingredients can be fed directly into the twin screw extruder, as well as very coarse ingredients.

96 What is the maximum level of fresh meat that can be incorporated in a recipe to be processed in a twin screw extruder?

Up to 30% fresh meat can be used in recipes processed on twin screw extruders.

97 What are the typical applications of twin screw extruders?

Originally developed for processing plastics, food companies began using twin screw extruders for products, like sticky caramels and candies, which could not be made with single screw machines. Very soon, twin screw extruders became popular with food manufacturers of specialized food items. Presently, twin screw extruders are being used for the following different food and feed items:

- **Snacks and cereals**
  - cereals and corn flakes
  - direct expanded snacks
  - third generation snacks
  - three-dimensional snacks
  - co-extruded snacks and other food items
  - multicolor food and snacks
  - corn chips and tortillas
  - chocolate filled snacks.
- **Bakery and pasta items**
  - precooked pasta
  - noodles, spaghetti and macaroni
  - topping and bakery analogs
  - bread-like products (e.g. crisp bread)
  - pastry dough
  - cookies and crackers
  - pretzels
  - instant noodles
  - ravioli.
- **Textured protein products**
  - texturized vegetable protein (soy)
  - wheat gluten textured products
  - texturized vegetable protein from partially defatted soy flour
  - meat analogs
  - high moisture meat analogs.
- **Feeds**
  - pet food
  - pet treats
  - aquaculture feed.

- **Bars and nuts**
  - meat and power bars
  - special energy bars with resin filling.
- **Miscellaneous foods**
  - food gums
  - marshmallow products
  - reformed fruit bits and sheets
  - imitation nuts
  - semi-moist food
  - soup and gravy mixes
  - instant rice
  - sugar crust liqueurs
  - three-dimensional confections and toffees
  - dairy products
  - cheese and casein products
  - beer powders
  - stabilization of rice bran
  - confectionery and other chocolate products
  - cocoa and crumb
  - egg rolls
  - jellies
  - flavoring.

98 What are the disadvantages of using twin screw extruders rather than single screw extruders?

Twin screw extruders are more expensive (at least double the price) than single screw extruders. Their maintenance costs may also be higher. They are relatively more complicated than the single screw extruder. Operators need to be very careful when assembling a new configuration. It is very easy to mount the segments in the wrong orientation causing them to break.

Selecting the right type of extruder: dry extruders and expanders for food and feed production

99 What is a dry extruder and what does dry extrusion mean?

A dry extruder is one that does not require an external source of heat or steam for injection or jacket heating. Dry extrusion capitalizes on the heat generated through friction, which is the sole source of heat for cooking and dehydrating the product. Sometimes dry extruders are referred to as autogenous extruders (referring to the fact they are friction-based), high shear extruders, single screw extruders, high temperature short time extruders (HTST) and low cost extruders (LCE).

100 What kinds of ingredients can be processed by a dry extruder?

Dry extruders can process ingredients which have a wide range of moisture contents, that is, 10–40%, depending on the premixed formula. If the ingredients have sufficiently low initial moisture content, drying the product after extrusion cooking may not be necessary.

101 How much moisture loss occurs in dry extrusion?

Moisture loss in dry extrusion is in the form of steam flash-off at the die and the extent depends on initial moisture in the ingredients and product exit temperature. Typically 5–10% moisture loss can occur during dry extrusion.

102 Can we inject water into a dry extruder?

Dry extruders have an option for water injection during extrusion. Usually starchy materials require some moisture in order to gelatinize.

103 Can a preconditioner be used with a dry extruder?

Yes, most dry extruders are equipped with preconditioners to enhance the yield and quality of the finished product.

104 What is the basic principle of dry extrusion?

Dry extruders are single screw extruders with screw segments and steamlocks (choke plates) on the shaft for increasing shear and creating heat. When material moves through the barrel and comes up against these restrictions, it is unable to pass through, pressure increases, and some back flow is created. Usually these restrictions are arranged in such a way that they increase in diameter toward the die end of the screw to create more pressure and shear as the product reaches the die. This build up of pressure and temperature, together with the shear stresses that have developed, plasticize the raw materials into viscous pastes or puffed shapes, depending on the raw material.

105 What temperatures and pressures can be reached in dry extrusion?

In dry extrusion, pressure and temperature are at their maximum just before leaving the die. The die design and opening also play very important roles in pressure build up. The cooking range in a dry extruder can be 82–160 °C (180–320 °F) with very high pressure (40 atmospheric pressure or, 587 psi). As soon as the material exits the extruder dies, pressure is instantaneously released from the products, causing the internal moisture to vaporize into steam, making the product expand and resulting in pasteurization of the product.

106 What are the main applications of dry extrusion?

A major use of the dry extruder is in preparing oilseeds for screw pressing of oil, primarily soybeans and cottonseed, although it has also been applied to sunflower, peanut and canola seed processing. In the process, soybeans and cottonseed are extruded using a dry extruder, then pressed in a parallel bar screw press to remove the oil. Extrusion prior to screw pressing greatly increases throughput of the expeller over the rated capacity. Oil and meal produced by this method are remarkably stable because extrusion also releases natural antioxidants in oilseeds. This process is used around the world for processing raw soybeans into full fat soybeans and partially defatted soybean meal. Dry extruders can also be used for processing cereal grains and other starchy materials such as tapioca and other tubers crops.

107 Are there any other applications of dry extrusion?

Yes, there are several other applications where a dry extruder can be used. Currently, dry extruders are being used for the following purposes in the food and feed industries:

- oil seed processing: including soybeans, canola, rape seed, cotton seed, sunflower seed and peanuts;
- complete feed: such as pig starters, aquatic feed (floating feed), dog food, cat food, mink and fox feed, poultry feed, horse feed and calf starter feed;
- processing high moisture by-products of the animal, food and marine industry into high quality feed ingredients;
- stabilization of rice bran through deactivation of the naturally occurring enzyme lipase;
- production of texturized plant protein and manufacture of cereal food;
- as a reactor to detoxify peanuts or cottonseed through reduction of aflatoxin from both seeds and reduction of the free gossypol in cotton seed;
- pasteurization and/or sterilization of feed and food;
- de-activation of antinutritional factors present in legumes;
- production of highly non-degradable (by-pass) protein for ruminant animals;
- production of slow release ammonia products for safe feeding of non-protein nitrogen to ruminant animals.

108 What are the pros and cons of dry extrusion?

Pros: Dry extruders require relatively low capital investment and can be engineered to fit all sizes and types of installation. Less training is required as compared with twin screw extrusion systems. Although there is the flexibility of adopting a steam preconditioner for applications such as complete feed or shaped products, dry extrusion is a particularly good choice when steam is not available. Dry extruders are able to grind whole soybeans during extrusion; therefore grinding steps can be eliminated in the case of soybeans.

Cons: Dry extruders require a relatively large amount of power to operate in some applications compared to other extrusion systems. Higher wear occurs on the screw because of the lack of lubrication from steam injection. Maintenance costs are also higher when whole soybeans are ground in the dry extrusion process. Grinding soybeans by a hammer mill beforehand may be more cost-efficient. They are not as flexible as wet segmented single and twin screw extruders. The final dimensions of the product may be limited to certain sizes; because of high pressure it is extremely difficult to shape products which are less than 2 mm in this type of extruder. The initial moisture content of the recipe is critical. Formulas with high fat contents may not cook properly because of product slippage inside the barrel. Highly viscous material is hard to process through these types of extruder compared to twin screw extruders.

109 What is the main difference between dry and wet extrusion?

There is no basic difference between the 'dry' and the 'wet' extrusion, except that more shears occur in dry extruders to create heat.

110 What are the pros and cons of wet extrusion compared to dry extrusion?

Wet extruders have a higher capital investment costs than 'dry extruders', but usually have lower operating costs. Wet extruders have higher capacities than dry extruders owing to the large drive motor requirements per unit throughput of dry extruders. 'Wet extruders' yield superior shaped products compared to 'dry extruders', owing to greater processing control.

111 What is an interrupted flight extruder?

The basic design of most of today's interrupted flight extruders was developed and introduced in the United States by the Anderson International Company (Cleveland Ohio) as the 'Anderson Grain Expander' in the latter 1950s for processing pet foods and other cereal products.

112 What is an expander?

Expander or interrupted flight expander are generic names in the oilseed industry for a type of cooking extruder.

113 How do interrupted flight extruders differ from other extruders?

An interrupted flight extruder (or expander) is mechanically different from other types of extruder because it was developed from the screw press. Screw presses and interrupted flight extruders are similar in that a revolving interrupted flight pushes the material through a cylindrical barrel and out through an opening at the barrel's end.

114 Do expanders have heating and cooling jackets?

Commercial interrupted flight expanders are not usually jacketed and rely on direct steam injection for supplemental heat beyond that created by the mechanical shear of the ingredients.

115 What is the main principle behind cooking in expanders?

An expander has a cooking effect due to the rapidly rotating worm shaft it contains, which has individual worms with interrupted flights. The worm shaft is positioned inside a smooth-walled barrel equipped with removable stationary pins which protrude from the barrel and intermesh with the interrupted flights. The purpose of intermingling the rotating worms with stationary pins is to provide a high shear, turbulent mixing action, which kneads the solid raw material with the injected water and steam. This ensures that there is rapid and uniform absorption of the injected moisture into all of the solid matter. As the steam is absorbed, it releases its heat of vaporization which elevates the temperature of the ingredients. Frictional

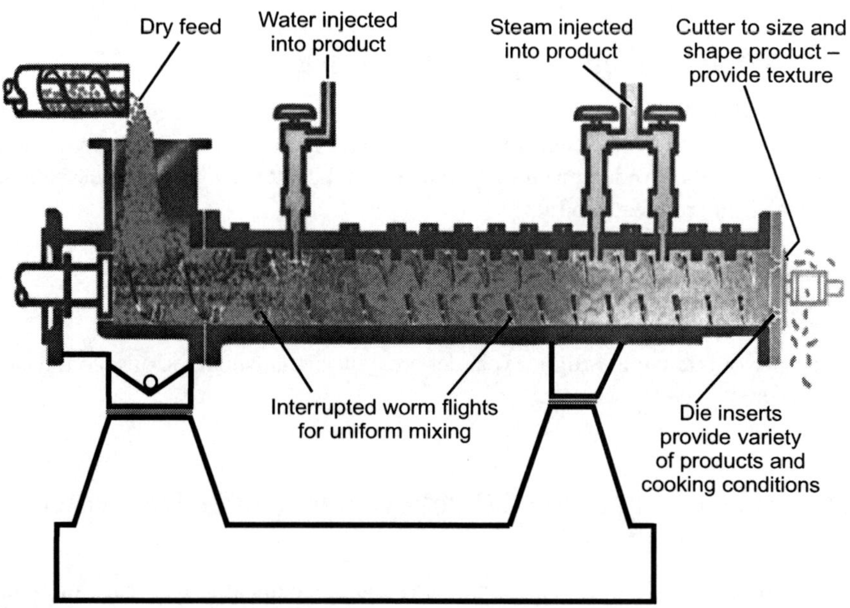

Fig. 1 A typical expander.

heat is also generated by the rapid motion of the flights, further elevating the temperature as they compact and work the mixture, subjecting it to increasingly higher pressures as it is forced through the length of the barrel. By the time the mixture reaches the end of the length of the barrel, it is thoroughly cooked and under sufficiently high pressure and temperature (120–150 °C; 248–302 °F) for much of the moisture to flash off as the product exits the expander. Fig. 1 shows a typical expander.

116 What are the main applications of expanders?

Expanders play a major role in preparing oilseeds for solvent extraction and special high shear heads have been developed for shearing oilseeds before screw pressing.

117 Can we remove oil from oilseed crops using expanders?

An oil removal cage can be added for high oil content seeds or oil-bearing materials to produce cohesive collets for solvent extraction and to increase oilseed traction in screw presses. Several screw ends and barrel head designs are available for processing different materials.

118 What is the particle size requirement for expanders?

An expander requires a finely ground or flaked material feedstock since they do not have the ability to grind themselves.

119 Can we add fat in an expander?

High levels of added fat can cause slippage inside the barrel and it is best to use ingredients containing internal fat when possible.

120 Can preconditioners be installed on expanders?

Yes, a wide variety of preconditioners can be adapted to be installed on expanders.

121 What are some other applications of expanders?

Expanders are being used for several other food, feed and oilseed crops. Some of the common applications of expanders are:

- Production of full fat soybean meal.
- Oilseed preparation for 'hard pressing' to 5–6% residual oil in the meal.

- Rice bran stabilization.
- Pet food production.
- Aquaculture feed production.
- Production of feeds for other animals (pig, cattle, horse, mink).
- Snack food production.
- Drying of synthetic rubber and plastic polymers.

122 What are the pros and cons of expanders?

Pros: Interrupted flight expanders are less expensive than segmented single- and twin-screw extruders. They are very simple machines which are easy to operate and minimum training is required for the operators. Interrupted flight expanders are also very rugged. As they have less shearing action than dry extruders less wear generally occurs in them. Smaller electric motors can be used, maintenance costs are lower and worn parts are easily replaceable. When properly installed and supplied with consistent ingredients, they produce a consistent quality product for long runs. Shaft speed and designs can be varied, which makes this machine applicable to different raw materials. Less power is required than for dry extruders of the same capacity because of lubrication from water (added as steam), but higher moisture products typically require drying after cutting.
Cons: Since steam locks are not present, heating in the barrel from shear is limited to heat resulting from the breaker bolts and the back pressure of the die plate and must be supplemented by injected steam. The maximum barrel temperature is limited by the open die area and by the pressure of the available steam and its quality. These machines are less versatile compared to segmented single- and twin-screw extruders and processing conditions are more difficult to control. Pellet sizes are limited to larger diameters compared to twin screw extruders. Since expanders have less shearing action than dry extruders, some products, for example soybeans, need to be ground before entering the machine, for example in the production of full-fat soybean meal.

123 What is a low cost extruder (LCE)?

A LCE is a low cost extruder used to process soybeans. Its applications are somewhat limited.

124 How do I know what type of extruder I need for my product?

Food and feed manufacturers who already use extrusion technology can answer this question very easily. They know the type of product and volume accepted by their market as well as any advantages their competitors may have in product appearance or functionality caused by machinery differences. However, for new companies or even experienced food manufacturers who are considering entry into a rapidly growing food and feed market, the decision is not easy. They have a harder time selecting the components and configuration needed. But, when the homework has been done, information gathered and the various factors considered, the number of practical alternatives should be refreshingly small. Often the final decision is based on local factors like cost and expected service from the extruder manufacturer.

125 What options are available to an extruder buyer?

Many options (which sometimes confuse buyers) are available in the marketplace when selecting extrusion systems for pet food, aquatic and livestock feed production. Questions that buyers face when selecting a system include:

- Should it be a 'wet' or 'dry' extruder?
- Is a single or twin screw extruder required?
- Should it be an expander with a pellet mill?
- Should it have continuous or interrupted flights?
- Should it have internal steam locks or a single-face die plate?

The options are numerous enough to guarantee confusion among first-time buyers. The fact that many extruder manufacturers are able to exist competitively indicates that most have found unique niches for their machines.

126 What are the considerations one should keep in mind when buying an extruder?

Questions to consider when selecting the appropriate extruder include the following:

- What will be the main ingredients used in the food, pet food, aquatic and livestock feed recipe?
- What is the physical nature of the main ingredients?
- What will be the physical and sensory properties of the food, pet food, aquatic and livestock feed?
- What are the moisture contents of the other main ingredients?
- What is the fat content of the ingredients that will be used in the food, pet food, aquatic or livestock feed formulation?
- What will be the availability of the constant or seasonal ingredients and substitute ingredients that may be used occasionally?
- What kind of food, pet food, aquatic and livestock feed will be made? For example in the feed industry, just a dry expanded pet food, or a semi-moist pet food or high-energy (extra fat) pet food, floating or sinking aquatic feed, high fat aquatic feed?

- Does the food, pet food, aquatic and livestock feed formulation contain high levels of fresh meat?
- Is the shape of the food, pet food, aquatic and livestock feed generic, exotic or detailed?
- Should the food, aquatic feed or pet food be multi-colored or center-filled with some other ingredients?
- What is the target bulk density of the food, pet food, aquatic and livestock feed to be made on the extruder?
- What is the desired production rate of the food, pet food, aquatic and livestock feed?
- The size of an extruder depends on the market size since extruders function best when operating at full throughput per hour.
- What will be the source of energy for the food, pet food, aquatic and livestock feed extruder? Is steam or electricity (for product heating) more economical where the extrusion plant will be built?
- For a small operation in a developing country would a tractor power take-off drive be more suitable?
- What are the capital availability and the recovery date target?
- Would a second-hand extruder fill the need better for a start-up operation?

Impact of protein, starch, fat and fiber on extruded foods and feeds

127 What is the impact of protein on extruded products in terms of expansion, texture and durability?

As the protein level in a recipe is increased – usually at the expense of starch – there are changes in product expansion, texture and durability. Higher protein levels usually result in less expansion as protein is less viscoelastic than starch. Increased levels of protein also result in firmer or harder textures compared to starchy recipes. Many proteins are not functional, in other words they will not contribute to expansion and binding in the extruded product matrix. Generally products made using recipes with higher protein levels will have less durability and may break and generate fine dust during handling and transportation.

128 How can I know if a proteinaceous ingredient will be functional during extrusion?

There are several tests that can be applied to a protein ingredient to determine its functionality. Functionality is defined as the level of water solubility which in turn affects the expansion and binding properties of the ingredient. Simply stirring a teaspoon of the ingredient into a cup of warm water can give a rough indication of functionality or water stability. If the sample quickly sinks to the bottom of the water and results in a semi-clear solution, the ingredient is not very soluble and will be likely to remain inert during the extrusion process. Figure 1 illustrates a relatively soluble and functional blood meal (sample on the left) and a low solubility (low functionality) blood meal on the right. A more quantitative test includes subjecting a sample to an official laboratory method recognized by

Fig. 1 Solubility comparison of animal protein samples.

several associations and referred to as PDI (protein dispersibility index). A test value of 100 PDI would indicate total solubility in water. A value of 0 PDI would indicate no functionality and no solubility in water.

129 What is a general guideline I can use to determine the functionality of a proteinaceous ingredient?

In addition to the simple test described earlier of stirring a sample in warm water and letting it set to indicate water solubility, the general color of a proteinaceous ingredient is an indication of PDI or functionality. Darker colors usually indicate poor solubility and functionality. Lighter colored proteinaceous ingredients usually indicate higher functionality and PDI values.

130 What are examples of proteins that are functional?

The functionality of a proteinaceous ingredient can be judged by the ingredient source. It is common for plant or vegetable proteins to be relatively soluble and functional. Examples of plant or vegetable proteins are soybean products, corn gluten, wheat gluten, and almost all the legume proteins and most cereal grains which contain some protein even though they are generally considered to be primarily starch sources. The functionality of this family of proteins can vary depending on historical processing temperatures during the manufacture, extraction, or refining of the ingredient. Ingredients exposed to higher temperatures during these processing steps may lose some functionality.

There are two groups of animal and marine proteins that also exhibit high functionality. Examples are proteinaceous ingredients that are subjected to mild processing steps such as spray drying. Even within the category of spray-dried proteins, however, there can be a wide variation in ingredient functionality due to the processing variables employed. Milk proteins, gelatin and some blood and plasma proteins would be included in this list. Animal and marine proteins that are injected in a raw or 'fresh' form are also very functional.

131 What are examples of proteins that are not functional?

Examples of proteins that are not functional usually include animal proteins. The large group of rendered products available from suppliers usually have very low functionality owing to the moderate or high temperatures and retention times employed during their manufacturing steps. Ingredient color and water solubility are simple tests to utilize in order to get a rough indication of functionality.

132 Do all proteins react the same in the extrusion process?

As discussed throughout the previous questions on this subject, there can be wide variation in protein functionality as well as nutritional value (Table 1). However, functionality and nutritional value do not necessarily correlate well. Proteins begin to lose their functionality when they reach temperatures of around 55 °C or higher. Loss of functionality simply means loss of water solubility and less contribution to expansion, binding and durability in the final extruded product. As previously mentioned, PDI values can give a quantitative indication of functionality. The temperature range at which proteins lose this functionality is sometimes referred to as the temperature of denaturation. On the other hand, proteins may not lose substantial nutritional value until they are subjected to temperatures of around 130 °C or higher. An indication of loss of nutritional value is the level of heat-damaged protein in the ingredient. Actual feeding trials are the best way to determine protein digestibility. It is possible to estimate the level of heat-damaged proteins quantitatively by measuring the amount of nitrogen in the acid detergent fiber fractions via wet chemistry methods. The relationship of process temperatures to protein solubility and heat-damaged protein is illustrated in Fig. 2. Note

Table 1 Protein biological value of common ingredients

Food	Biological value
Egg	100
Fish meal	92
Milk	92
Liver	79
Beef	78
Soybean meal	67
Meat and bone meal	50
Whole wheat	48
Whole corn	45

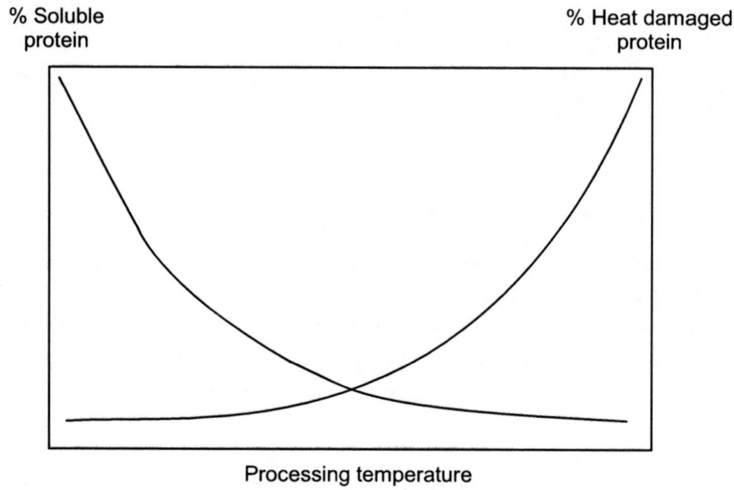

Fig. 2 Soluble and heat-damaged proteins versus process temperatures.

that no temperature values are listed as proteins vary somewhat in their response to process temperatures. As a guideline, the two curves cross at around 150 °C.

133 At what temperatures do proteins lose their functionality?

Proteins typically begin to lose their functionality, or become denatured, in the 55–65 °C range. The exact temperature at which this occurs depends upon several factors including moisture content, pH and pressure.

134 What does starch contribute to product attributes during extrusion?

The contribution of starch to product expansion and binding or durability is much greater than that of denatured proteins and somewhat better than that of functional proteins. Starch is unique in that its granules begin to absorb water and swell as it approaches gelatinization temperatures (60–75 °C). When heated in the presence of water, starch granules eventually reach a point where the swelling is irreversible, the native crystalline structure becomes amorphous and the total starch more soluble. Native starch generally refers to raw unmodified starch. Amylose can leach out of the granules and contribute to extrudate elasticity and viscosity. It is these properties that improve binding and assist in expansion at the extruder die. If a lighter density product is desired, increasing the starch level will improve expansion. If a stronger, more durable product is desired, increasing the starch level will improve binding to advance these properties.

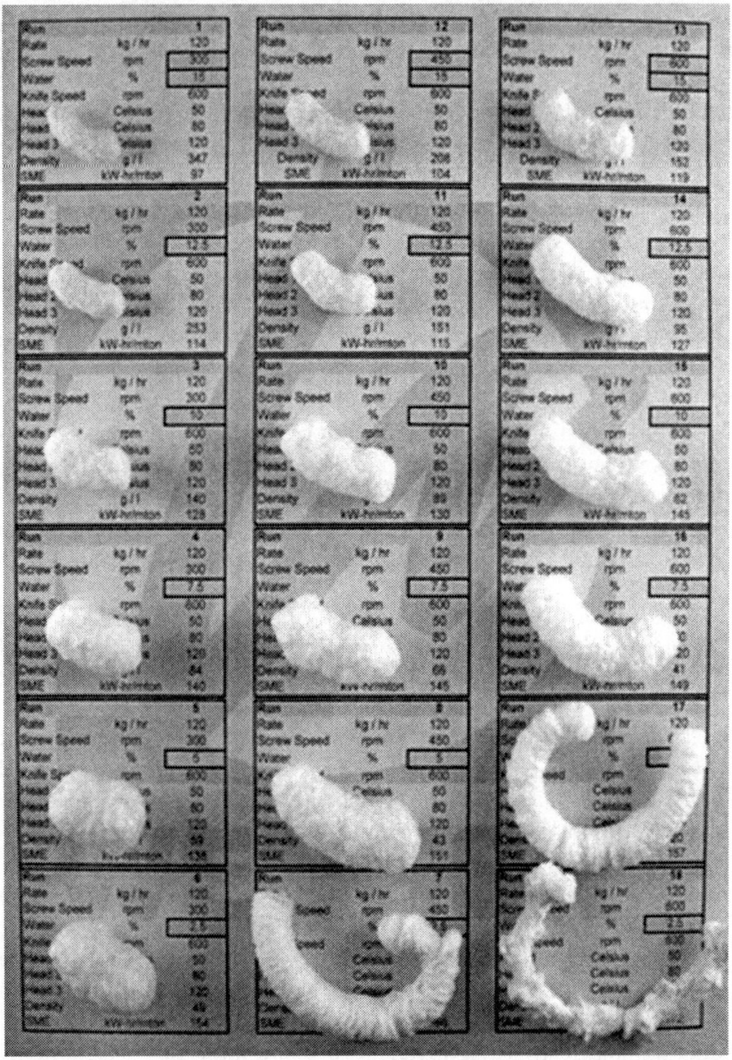

Fig. 3 Impact of screw speed and extrusion moisture on extrudates containing high levels of starch.

Common sources of starch are food grains including corn, wheat, rice, oats, barley and sorghum, and common root or tuber crops including potatoes, sweet potatoes, yams and cassava (tapioca). Many legumes are also a major source of starch and are becoming more common ingredients in extruded recipes, often driven by human food marketing initiatives or by food safety concerns.

As a general rule, gelatinization of starches is greater at higher moisture content as long as sufficient energy is present during extrusion to break intermolecular bonding. Shear can be increased either by changing the screw configuration or by increasing the screw speed. Figure 3 illustrates the impact of screw speed and extrusion moisture on a directly expanded snack made from degerminated corn meal (78% starch).

Under normal extruder conditions (high temperature, high shear) in the production of many food products, native starches in grains are preferred. In general, the following statements can be made regarding extrusion and traditional grains: (1) increasing extruder barrel temperature will usually increase expansion of extruded starches, (2) increasing the percentage of amylose (up to about 50%) will usually increase expansion, (3) screw speed alone can influence expansion, (4) changing the screw configuration to increase friction and shear may result in greater expansion and (5) lower moisture tends to increase friction and shear forces in the extruder barrel and results in increased expansion. Published papers clearly demonstrate that conditions for optimal gelatinization and expansion are highly dependent on processing variables and will change for any given combination of ingredients. Extrusion processing of most foods at less than 20% moisture may result in increased screw and barrel wear as well as increased opportunity for the production of dextrins. Dextrins are produced under high temperature–high shear conditions when there is not enough water to hydrate the starch adequately.

135 What levels of starch are typically found in extruded products?

Typical starch levels in pet foods cover a broad range of inclusion. Several years ago the norm for starch inclusion was 20–50% and most preferably in the 30–40% range. In the last few years starch has reached levels as high as 65% and as low as zero (no starch inclusion). The higher levels of starch are more common in reducing diets where fat and protein levels are reduced and starch and fiber levels are increased. There has also been tremendous interest in low carbohydrate and 'no carbohydrate' diets which have pushed starch inclusion rates downward, even to the point of not being included in the recipe. The low starch inclusion rates have made product durability a challenge which has been met to some extent by the increased levels of functional proteins in the recipe.

Floating aquatic feeds generally require a minimum of 20% starch to create sufficient expansion and low densities for high buoyancy properties. Sinking aquatic feeds can have starch levels as low as 10% where the starch mainly serves as a binding agent in the recipe.

Snacks, cereals and other extruded foods usually have a substantial amount of starch and occasionally a recipe containing 100% starch is extruded to create a cold-water soluble product.

136 Do different starch sources react differently in the extruder?

The source of starch can have a significant impact on the extrusion process.

Table 2 Heat of gelatinization of various starches

Starch source	Heat of gelatinization (cal g^{-1})	Amylose content (%)	Granule size (μm)
High amylose corn	7.6	55	5–25
Potato	6.6	20	15–121
Tapioca	5.5	22	5–35
Wheat	4.7	28	1–35
Waxy corn	4.7	0	5–25

Table 2 lists some common starches and the differences in granule size and energy required for gelatinization. Rice starch has very small granules that are tightly packed together and which hydrate slowly. However, once the hydration has progressed and gelatinization occurs, rice starch can become very sticky. Long and medium grain rice varieties will be much less sticky during the hydration steps of preconditioning and extrusion compared to short grain varieties. Rice starch also has significant digestibility in most gastrointestinal tracts even when gelatinization values are modest. Corn starch is a good binder and at levels above 40% in the recipe, the extrudate can also become very sticky. The wheat grain is unique in that in addition to the significant levels of starch present, there is also a functional gluten protein available which is an excellent binder. The tuber starches of potato and tapioca (cassava) are excellent binders even at levels as low as 5%. These starch sources will often result in a smoother surface on the extruded product.

137 What is the impact of adding more starch to a recipe?

As starch levels are increased in a recipe, the resultant extrudate will generally expand more yielding lighter bulk densities. Increased starch levels will also contribute to product binding and durability. Above 65% levels, starch may create some challenges during extrusion as it can become very sticky. Increased starch levels can also have an impact on the ability of an extruded and dried product to absorb topical liquids during the enrobing and coating steps.

138 Is there a minimum moisture level required during the extrusion of starch?

Traditional experience has indicated that extrusion moisture needs to be at least 30% to realize maximum gelatinization during extrusion. Testing outside of the extrusion environment with other techniques such as the DSC (differential scanning calorimeter) method has confirmed this assumption (see Table 3).

Table 3 Minimum moisture levels required to initiate starch gelatinization

Starch source	Moisture (%)
Wheat	31
Corn	31
Waxy corn	28
High amylose corn	34

139 How do I measure how well starch is cooked or gelatinized during extrusion?

There are a number of methods that have been used to measure starch gelatinization. The methods include using an RVA (rapid visco analyzer), a DSC (differential scanning calorimeter), hydration and water absorption techniques, NIR (near infrared) and enzyme susceptibility methods. There appears to be several important advantages to the enzyme susceptibility test:

- It more closely mimics actual digestibility results as both treatments use enzymatic hydrolysis to convert starch to glucose.
- The sample size is large enough to give more representative results.
- There appears to be little or no interference from other recipe components.
- No base or raw sample is required for each analysis as each sample is analyzed for total starch and gelatinized starch.
- This test has consistently shown acceptable results over 30 years of use.

140 What are soft and hard types of grains and what impact does this have on the extrusion process?

Many grains contain small starch granules that are tightly packed together resulting in a very hard grain that hydrates slowly and is more difficult to cook during the extrusion process. Soft grains usually hydrate more quickly and thus require lower energy input to accomplish the desired level of cook. Even within the grain kernel there are hard (horny endosperm) and soft (floury endosperm) regions. If grains are milled and sifted to separate these regions, flours with vastly different extrusion properties can be manufactured.

141 How does fat affect the extrusion process?

Fat is routinely added to a recipe as it is processed through the extruder to provide additional calories, to improve palatability and acceptance, to provide nutrients such as fatty acids, as a carrier for fat-soluble vitamins, or for dust control. Fat

Table 4 Effect of internal fat levels on expansion of extruded products

Added fat to extruder (%)	Extrudate bulk density ($g\ l^{-1}$)
0	253
5	308
10	408
15	528

addition also has an impact on the extrusion process as fat acts as a lubricant which can reduce friction between particles in the recipe and between the recipe and the internal components of the extruder. The lower the amount of friction, the lower the mechanical energy input required, resulting in less expansion (greater product bulk densities) as shown in Table 4.

Fat also disrupts the product's internal matrix and thus weakens the overall structure resulting in decreased product durability. At inclusion rates of less than 7%, there is usually a minimal effect on product characteristics. When addition levels reach the 7–12% range, the product density begins to increase even when efforts are made to maintain bulk density. When fat addition levels are 12–17%, product may have little or no expansion, but will retain some durability. At levels exceeding 17%, durability may be significantly impaired.

142 Is there a difference in impact between different fat sources?

Fat or oil sources (all from the general category of lipids) can be from many sources. The most common are the well-known vegetable oils and animal or poultry fats. Marine oils are also used and many times combinations of all these sources are utilized to achieve the desired fatty acid profile for nutritional purposes. There appears to be little or no difference in the impact of the fat on the extrusion process as far as the fat source is concerned.

143 Why is fat added to a recipe during extrusion?

As mentioned previously, fat can be added for nutritional purposes as well as a means to drive the extrusion process in different directions.

144 What other effects do fat levels have on the extrusion process?

As fats levels are increased in a recipe during extrusion, a larger cell structure will result in the final product. The cells will also have thicker walls. This has the impact of slowing hydration if the final product is placed in water. The thicker cell walls also reduce uptake during the coating process if the product is to be enrobed. Products containing more internal fat also appear to be more difficult to dry, perhaps due to thicker cell walls and the slightly higher extrusion moisture employed during thermal processing.

145 How can I maximize the level of fat in an extruded product?

There are several steps that can be taken to maximize the level of fat in an extruded product. The following is a list of steps that can be taken:

- Formulate the recipe with ingredients that are high in indigenous fats (examples would be full fat soy and flax meal). The same level of fat can be achieved in the final product with less impact on extrusion if the fat is supplied as part of an ingredient rather than added in pure liquid form.
- If fats must be added in liquid form, it is advantageous to heat them to 40–60 °C just before extrusion to improve the efficiency of the thermal cooking process. Nearly 90% of the energy inputs in an extrusion process are consumed in raising the recipe constituents to the cooking temperature. Any steps that can be taken to 'preheat' the individual recipe components will improve the cooking efficiency.
- Fats should be added as late as possible in the process to reduce their impact on extrusion. If fats can be added in the coating steps, there is no impact on extrusion.
- As internal fat levels are increased in the recipe during extrusion, it will also be necessary to increase starch or functional protein levels to maintain expansion and binding properties.
- Fat brings lubricity. It will be necessary to increase the thermal or mechanical energy inputs to maintain product attributes such as the desired density and durability.
- It is often helpful to increase extrusion moisture during processing of high fat recipes as water is a necessary solvent for starch gelatinization. On many occasions operators are inclined to reduce extrusion moisture as the fat gives the extrudate a deceptively wet, shiny appearance as if it is high in moisture.

146 What is the impact of fiber on extrusion?

At inclusion levels of less than 5%, there is very little impact of fiber on the extrusion process. This is especially true if the particle size of the fibrous ingredient is less than 400 μm. In general, it appears that a smaller particle size has a less detrimental effect on expansion as a size of less than 50 μm yields a very fine cell structure in the extruded product. Very large or coarse particles in the fiber source will limit expansion and can result in a very rough surface appearance. The more soluble forms of fiber have less impact on expansion. There are indications that extrusion could slightly increase soluble fiber levels as a result of extreme processing conditions.

147 Are there different types of fiber?

Fiber sources can be conveniently divided into two groups based on their solubility and fermentability. Soluble fibers are generally very fermentable. Examples of soluble fibers are beta-glucans from oats and barely, fruit pectins, psyllium seed, inulin, root vegetables, legumes and some gums. The more insoluble fibers are not very fermentable and include whole grain bran, some vegetables such as celery and zucchini, fruit skins, vegetable peelings and resistant starches. Insoluble fibers are inert during extrusion and do not contribute a great deal to expansion and binding. Fermentable fibers, which are more soluble, can improve expansion and binding of extrudates.

148 Is there a critical particle size for fibrous ingredients?

The standard guidelines for recipe particle size apply for soluble or fermentable fibers. However, insoluble fiber sources should have a particle size of less than 400 μm or the final product could have a course appearance and rough surface. As discussed above, insoluble fiber that has a very fine particle size (less than 50 μm) will actually yield an extrudate with a very fine cell structure, as the insoluble fiber particulates can serve as nucleating sites for cell development.

149 Should I be concerned about acrylamide in extruded products?

Acrylamide is formed in high-carbohydrate foods during high temperature processes such as frying, toasting or extrusion. Acrylamide is known to form in home-cooked foods, mainly potato and grain-based products. Extreme processing conditions, often accompanied by color development, are indicators. Frying steps of 150–190 °C for 10 minutes or dark toasting of toasted bread slices may result in the formation of acrylamide. The high temperature but short term process of extrusion results in low to negligible levels of acrylamide.

150 Should I be concerned about biogenic amines in extruded products?

Biogenic amines are products of bacterial fermentation that can occur in some animal proteins when stored under improper conditions. Biogenic amines may already be present in the raw materials coming into the extrusion plant. Extrusion will destroy the bacteria responsible for their formation but if the biogenic amines are already present, extrusion is not effective in reducing their level. The best prevention is to monitor raw materials for their presence and ask for a certificate of analysis from the supplier stating that they are present in levels of less than 300 ppb.

Impact of particle size and other ingredients on extruded foods and feeds

151 What is the proper grind or particle size of raw materials for extrusion?

Choosing the proper particle size or grind prior to extrusion is dependent on several factors. The size of the die orifice, the cell wall thickness and cell or pore size desired in the final product, and the surface appearance are all greatly influenced by the particle size of the recipe prior to extrusion. Small particles in the recipe prior to extrusion usually indicate that this raw material will hydrate and cook more easily.

It is most common for recipes to contain particles that are too large rather than particles that are too small. However, grinding a recipe to a very small particle size can result in very high grinding costs as recipe throughput will be greatly reduced in the milling operation. If all particles are less than 400 μm in some recipes, the conveying properties of that recipe will be compromised and the smaller particles may actually hydrate too quickly as they preferentially absorb the available water. The final result can be non-uniform hydration and sticky properties which could affect the stability of the extrusion process. A typical particle size distribution of a recipe ground through a 1.2 mm opening in the hammer mill screen is shown in Table 1. It is often the practice to grind whole grains through a 4.5 mm screen opening as the commodity is received. The coarsely ground grain is then mixed with other meals and ingredients to arrive at a complete recipe before the final milling step.

152 What are the benefits of having raw materials with particles of the proper size?

Choosing the proper particle size for the raw recipe can have many benefits. Benefits include the following:

Table 1 Typical particle size analysis of a recipe ground through 1.2 mm hammer mill screen openings

US standard sieve	Openings size	Percentage on sieve (%)
20	850	1.00
30	600	2.00
40	425	28.00
50	300	29.00
60	250	21.00
80	180	14.00
Pan	0	4.00

- Improved product appearance.
- Reduced incidence of die orifices plugging.
- Greater ease of cooking
  - reduced product breakage and fines
  - increased water stability and absorption
  - lower bulk density and softer texture
  - higher levels of starch gelatinization
  - higher throughputs for a given extruder and preconditioner
  - less wear on extruder components for a given rate
  - lower microbial survival rates
  - improved digestibility and palatability.
- Improved retention of liquid coatings owing to smaller cell structure size.

153 What do consumers perceive when extruded products are made with improper size particles?

When the particle size of the raw recipe is evident in the final product, the consumers' perception is that the quality of the final product is compromised. This

Fig. 1 Extruded product from a recipe ground through two different screen openings in the hammer mill.

perception can be formed from the obvious presence of large, visible particulates on the surface of the extruded product. The particulates could be large pieces of grain or other ingredients such as bone fragments. Actual quality in terms of nutrition may not be affected but the perception is critical to the marketability of a product. An example of the impact of proper recipe particle size on product appearance can be seen in Fig. 1. The sample product on the right was extruded from a recipe ground through 1.5 mm openings in the hammer mill screen. The sample product on the left was extruded from a recipe ground through 1.0 mm openings in the hammer mill screen. Both products were processed through the same size die orifice and operating parameters. The only difference was the particle size of the recipe prior to extrusion.

154 What guidelines can I use to determine the proper particle size of raw materials?

There are guidelines that can be used to determine the proper particle size of a recipe prior to extrusion. The first segment of the guideline is that the largest particle should be less than one-third of the diameter of the final die orifice. If there are particles in the recipe that are larger than one-third of the die openings, then it increases the possibility that the die orifices will be plugged or partially plugged resulting in reduced throughputs, poor or distorted shapes, and eventually shutdown. The second segment of the guideline states that no particles should exceed 1.5 mm in diameter. For example, if the die orifice is 9 mm diameter, the largest particle size should be 3 mm in diameter so one would choose a hammer mill screen with openings if 3 mm or less. However, because no particle should exceed 1.5 mm, it would be necessary actually to choose a hammer mill screen with 1.5 mm openings.

Using another example, if the die orifice is 3 mm in diameter, then one-third of this is 1 mm as the choice for the openings in the screen of the hammer mill screen. This is smaller than the segment of the guideline that states no particles should exceed 1.5 mm in diameter, so the proper screen opening in the hammer mill would be 1 mm.

It would be impractical to change the screen size in the hammer mill for every change in the final die orifice size. It may be more practical to divide all the products into one or more families based on the size of the final die orifice. Many manufacturers of extruded products simply choose a hammer mill screen that is dictated by applying the guidelines to the smallest diameter die opening and keep this screen installed for all products.

155 What is the impact of using rework in a recipe to be processed by extrusion?

Fines, broken product and under-processed material make up a stream of material that can be collected throughout the process and reworked back into it. This stream of material is known as rework. Usually the rework is dried and ground to the same particle size as the raw recipe and blended using the same recipe at 5–10% levels. It is important to maintain a fairly consistent level of rework to avoid changing the process parameters or final product characteristics. If a process fluctuates from no rework inclusion to inclusion rates of 20%, the process and final product could change drastically. As the level of rework is increased in the recipe, the final product will become darker in color, exhibit less expansion and higher bulk densities, have higher levels of cook and have a more defined shape.

156 How can a recipe be adjusted to make it shelf stable after extrusion and cooling?

Shelf stability usually refers to the ability of a final product to maintain product characteristics without becoming rancid or growing mold. The ability of a product to support mold growth depends on the water activity of the product. Microbiologists generally agree that water activities of less than 0.70 will be stable and not support mold growth. When the final product is dried to the 10–12% moisture range and the recipe pH has not been adjusted, a target water activity is usually around the 0.62 mark, as water activity is the only barrier to mold growth.

If 'hurdle technology' is employed, the water activity may be allowed to increase to near 0.7. 'Hurdle technology' refers to the practice of employing more than one obstacle to mold growth such as controlling water activity and product acidity or pH. This technique is commonly used for semi-moist products where moisture levels can be as high as 30%. The water activity is controlled to 0.7 through the addition of humectants to the recipe along with pH adjustment with acidulates in the 4.0–4.5 range, and with the inclusion of mold inhibitors. This preservation system (with the hurdle technology) is common in many high moisture food and feed products.

Examples of common humectants are glycerin, corn-syrup solids and propylene glycol. Acidulates in liquid form include phosphoric acid and lactic acid. Solid acidulates include sodium bisulfate. Almost any food with moisture levels above 10% will require a preservation system or must be coupled to low temperature storage (refrigeration and freezing).

157 Which colors or pigments are used in extruded products?

Dyes and pigments can be added to extruded recipes to impart a desired color or hue. Synthetic or FD&C (food , drug and cosmetic) colors are certified in the USA and include any one or blends of the following: Blue #1, Blue #2, Green #3, Yellow #5, Yellow #6, Red #40, and Red #3. Blue #1 and Green #3 are not broken down in the body when a product is consumed and are then excreted. There are also many uncertified or natural colors and pigments which are very popular in many food and feed products. A partial list of these colors and comments are found in Table 2.

Table 2 Uncertified or natural colors and pigments

Color or pigment	Comment
Caramel	Brown, red color
Iron oxides	Yellow, red, or brown color
Titanium oxide	White pigment limited to 1% maximum inclusion rate
Carmine	Red color for normal pH
Anthocyanins	pH sensitive
Carotenoids	Yellow, red or orange
Annatto	Orange
Paprika	Orange
Turmeric	Yellow
Beetroot	Red to purple color
Chlorophyll	Green
Lutein	Yellow to orange-red

158 Which natural pigments or colors are suitable for use in extruded products?

A list of natural colors and pigments include the following: caramel (browns), iron oxides (reds), carmine (reds and oranges where normal pH is the environment), anthocyanins (reds and oranges but are pH sensitive), annatto (orange), paprika (orange), tumeric (yellow), alfalfa (green), beet (red to purple), chlorophyll (green) and corn gluten meal (yellow).

159 What can be added to the recipe to prevent fat rancidity?

Rancidity can occur in many products or ingredients during storage . It affects taste and odor, and can have an impact on nutritive value. Two major rancidification pathways are recognized in extruded products. Enzymatic rancidity is catalyzed by the presence of certain enzymes (examples are peroxidases and lipases). The heat treatment experienced during extrusion is usually sufficient to destroy these enzymes. However, if ingredients containing these enzymes are stored in their raw or untreated state, they will become rancid before extrusion and will contain rancidity by-products that will impact flavor and odor. For many ingredients (especially whole seeds), it is important to heat treat the ingredient as quickly as possible after the seed coat is broken as this action triggers the enzymes responsible for rancidity. Extrusion is very effective in controlling enzymatic rancidity if the raw materials are processed immediately after milling.

The second pathway responsible for rancidity is oxidation. Oxidative rancidity is a chemical reaction catalyzed by heat, ultraviolet light, heavy metals and oxygen. Many of these factors are obviously present during the extrusion process. Oxidative rancidity is best controlled by the addition of antioxidants and 'oxygen scavengers'. Most antioxidants are added before extrusion even though there may

Table 3 Retention of vitamin activity under typical extrusion conditions

Vitamin	Retained (%)
A	80–88
D_3	Stabler
E	85
Riboflavin	74–100
Pantothenate	90
Niacin	80
Choline	Stable
Biotin	90
Thiamine	88
Ascorbic acid polyphosphate	96–100

be minor destruction during the process depending on the time and temperatures employed. There is a growing trend to utilize natural antioxidants such as mixed tocopherols, rosemary, and other unique ingredients. Often an 'antioxidant package' containing antioxidants and oxygen scavengers are used in synergy. Oxygen scavengers such as ascorbic and citric acid appear to increase the effectiveness of antioxidants.

160 What is the effect of extrusion on vitamin stability?

Vitamins are required in our diets and are often added to extruded products. Although extrusion is often carried out at relatively high temperatures, the retention time is very short and thus there is less destruction than expected. There are many factors that have an impact on vitamin retention during extrusion including extrusion moisture and vitamin concentrations in the recipe. Table 3 is a summary of typical vitamin activities retained during extrusion.

161 What is the effect of extrusion on minerals?

Extrusion usually has little impact on minerals as they are fairly inert. Extrusion processing may reduce mineral particle size and embed the minerals throughout the product matrix owing to the dispersive and distributive mixing forces present. Minerals, on the other hand, can accelerate wear on extruder components owing to their abrasiveness.

162 Will extrusion control undesirable components in my recipes?

The extrusion process is a very effective thermal treatment for controlling undesirable heat labile components. Examples of heat-sensitive components that can be reduced in content, totally destroyed or inactivated during extrusion include the following:

- ANF (anti-nutritional factors) such as trypsin inhibitors.
- Enzymes.
- Microbial populations.
- Viruses.
- Insects.
- Toxins.

The moist heat that is utilized in most extrusion processes is especially effective.

163 What flavorings are used in extruded products?

Oil- or water-based flavorings are much more stable during extrusion than alcohol-based or other volatile flavoring compounds. Flavorings that are salt based such as monosodium glutamate are also effective.

164 What impact does extrusion have on natural flavorings?

Extrusion will greatly reduce the impact of volatile natural flavors. As stated in Question 163, flavors that are not volatile are the flavorings of choice. Extrusion can also develop certain flavors if desired. A 'cooked cereal' flavor or 'toasted' flavor is possible with the correct recipes and processing conditions.

165 Do raw materials (other than flavorings and colorings) contribute any flavor or color to the final products?

Raw materials can make a huge contribution to the flavor or color of the final product. Examples are corn-based products (which contribute a yellow color) and many of the plant-based ingredients where chlorophyll and other compounds contribute to color development. As extrusion temperatures increase, the color development may be altered or trend towards browning reactions. A well-known example of a browning reaction is the non-enzymatic Maillard reaction which is especially prevalent when proteins and reducing sugars are present. In general, as product expansion increases, color development will be less intense as the 'puffing' tends to fade the color.

166 What impact do calcium carbonate and other similar ingredients have on the extrusion process?

Most minerals can have a buffering effect on the pH of the product or can actually move the product pH to the basic or acidic direction as desired. Calcium carbonate present as very small particles will serve as nucleating sites in the expansion process at the extruder die, resulting in small uniform cell structure. This phenomenon can alter product appearance and texture. Many additives to the recipe will also have an impact on flavor. For example, calcium carbonate, calcium hydroxide, or calcium oxide (lime) can impart a 'masa' flavor to extruder corn products.

167 What is the impact of adding salt or sugar to a recipe?

The obvious impact is a flavor contribution on the salty or sweet side depending on which ingredient is used and the level of incorporation. Salt, in combination with certain proteins found in meat products, will extract myosin from the muscle proteins during extrusion resulting in a strengthening of the product matrix and structure. Sugar in a plasticizer and will reduce the glass and melt transition temperatures of a recipe during extrusion. Sugar will actually liquefy during the heat and shear of the extrusion process reducing extrudate viscosity. Upon cooling, the product will become more crisp.

168 Why are encapsulated vitamins sometimes used in extrusion? Is encapsulation of vitamins effective?

Encapsulated vitamins are sometimes used to improve the retention of vitamins during storage and processing. Encapsulation is more effective as a protective measure during storage as the extrusion process has significant shear inputs which degrade the protective coatings or matrix used in the encapsulation process.

169 Which extruded products are likely to become rancid if antioxidants and oxygen scavengers are not used?

Extrusion promotes oxidative rancidity owing to the presence of heat and oxygen during processing and thus all extruded products are susceptible.

170 How do I choose which antioxidant and oxygen scavenger to use and at what levels to add them?

It is advisable to follow the manufactures' recommendations when choosing which additives to use and at what level.

171 In what forms are raw materials typically purchased?

For food extrusion, cereal grains are usually purchased as dehulled and degerminated with the bran also removed. However, in many circumstances, whole grains are utilized but are milled to small particle sizes to assist in processing. Tuber products are usually purchased as flours but meals and flakes are also available.

Preconditioners in food and feed extrusion: common problems and their solutions

172 Why doesn't my preconditioner cook or precondition my product sufficiently?

The function of a preconditioner is to 'precondition' a recipe prior to extrusion. Preconditioning involves the three main principles of hydration, heating and mixing. All three of these processes can be affected by the retention time in the preconditioner: if this is too low, the amount of hydration, heating, and mixing that occurs can be reduced. Hydration can be increased by increasing the level of mixing, the retention time and the amount of moisture added in the form of steam or water. Heating can be increased by adding more steam or increasing the temperature of the added water. Mixing can be increased by increasing the retention time and/or the mixing intensity of the preconditioner. Use of double-shafted preconditioners will generally increase the level of mixing. These have proved to be the most efficient type of mixer.

Cooking or gelatinization of the starchy portion of the recipe also occurs in the preconditioner and is a function of the same parameters described above. Gelatinization of starch requires heat, hydration and retention time, thus increasing preconditioning will result in an increased cook. It is recommended that at least one-third of the gelatinization should occur in the preconditioning step to improve the efficiency of the cooking extrusion step. If a preconditioner fails to cook the raw materials properly, this is because it is not hydrating, heating and/or mixing them correctly.

173 What can I do to increase the retention time in my preconditioner?

Retention time is a function of the mass of material in the preconditioner at any given moment divided by the rate of material throughput. Thus in theory the retention time in any preconditioner can be increased by increasing the fill level of the preconditioner or by reducing the throughput. Most preconditioning devices, though, have little or no control over retention time. The ability to vary shaft speed is the most common method of controlling throughput and thus retention time, but if the preconditioner is not equipped with a variable speed drive, the shaft speed cannot be changed. The pitch or angle of the beaters are also sometimes designed to be repositioned so that they retain more material in the preconditioner, which has an impact on fill level and thus changes retention times. This is not an on-line adjustment: the system needs to be stopped to make the necessary changes to beater orientation. Some preconditioner designs have a restriction device incorporated in the preconditioner's discharge opening which can control fill levels. Another alternative is to install a larger volume preconditioner which will retain more mass at any given moment at a given recipe throughput, resulting in longer preconditioner retention times.

174 Why does my preconditioner become plugged up at the discharge?

A preconditioner will usually become blocked at the discharge opening when the material being preconditioned forms large chunks which physically block the opening or when the material is so sticky that it gradually builds up on any restricted opening and flow is eventually blocked off. Raw materials exhibit many different properties depending on the nature of the ingredients in the recipe. Those that become sticky when heated or hydrated are more susceptible to this type of blockage. On most occasions this situation can be avoided by simply reducing the amount of steam and/or water added to the preconditioner.

175 How can I reduce the particle size of product coming out of my preconditioner?

As discussed above, the particle size of the material discharging from the preconditioner can be reduced by reducing the moisture and/or heat of the recipe in the preconditioner (see Fig. 1). Increasing the mixing intensity will also reduce the size of the chunks. Again, the mixing intensity can be improved by moving to a double shaft design of preconditioner.

Fig. 1 Preconditioner sample (a) from low mixing intensity process, (b) from high intensity preconditioner process.

176 How can I add higher levels of liquids and slurries to my preconditioner without plugging up the preconditioner at the discharge?

Most preconditioning systems are employed to blend or incorporate various liquids and slurries into the dry recipe on a continuous basis. As the level of these liquid components is increased, the resulting blend becomes less free-flowing and stickier. This can, at some point, result in the formation of large clumps of material that will block the discharge of the preconditioner. Liquids vary in their stickiness and viscosity and this will determine the maximum levels that can be added without plugging problems occurring. Increasing the mixing intensity may allow higher levels of liquids to be added. The viscosity of many liquids is temperature dependent and thus heating liquids prior to injection into the preconditioner may reduce viscosity and stickiness and allow higher levels to be added.

177 Why do my steam injection lines to the preconditioner become plugged?

The plumbing lines used to convey steam to the preconditioner can occasionally become plugged with material. Most generally this occurs during the shutdown mode when the steam flow is turned off before material has emptied out of the preconditioner. It is possible to purge steam lines of material by simply turning on steam flow to the steam injection lines for 3–5 seconds after the preconditioner has been stopped. Some plumbing designs install a check valve at the end of the steam injection line to prevent substantial accumulations of product in the steam injection lines. Plumbing lines can be designed to allow easy and quick removal of material that may have accumulated in them. For example, it may be possible to remove a plug to allow a wire or tape to be inserted into the line and this can be used partially to clear it.

178 Why do the water spray nozzles become plugged?

Water is introduced into the preconditioner through a series of lines that often include spray nozzles at the very tip or an injection point where water is discharged into the preconditioner. Spray nozzles greatly improve the distribution of water into the material by creating a mist or a fan-shaped spray pattern that allows increased contact of water with the product. Spray nozzle orifices are best sized by selecting a nozzle that will deliver the spray pattern at the flow rate utilized when water pressure is 40 psig (2.7 bar). The orifices at the tip of the nozzles can become plugged if they are not being used and preconditioner product accumulates around them and hardens as it dries after the system has been stopped. Nozzle tips can also be plugged by debris in the water. This debris can come from the water itself or consist of particulates such as rust that may flake off into the water stream over time. The plugging of nozzles by debris in the water can best be stopped by inserting a filter or strainer in the water line just upstream of the nozzles. The filter arrangement should be such that it can easily be cleaned while in operation or a second filter can be installed in parallel and used while the primary filter is being cleaned and serviced.

179 Why does the preconditioner drive stall or overload during startup?

The drive motor on a preconditioner is properly sized by the manufacturer to turn the shaft(s) of the preconditioner easily during normal operation. Occasionally the preconditioner components are adjusted by an operator in an effort to increase retention time which in turn increases fill. If a preconditioning system is stopped suddenly without allowing its contents to partially empty, it is possible that the drive may overload during a subsequent startup. This may also happen if the

Fig. 2 A preconditioner that is overfilled.

material is sticky and hardens during the time the system is out of operation or if the preconditioner fill level is excessive (see Fig. 2). The preconditioner manufacturer may advise a larger drive or a modification of shutdown procedures if this event occurs continually.

180 Why is there excess material residue in the preconditioner?

Excess material residue may occur for the reasons discussed in Question 179. Residue can also be formed because of an excessive buildup of material on the inside walls, beaters and shafts inside the preconditioner. This can occur during normal operation if the recipe contains components that are very sticky when heated or hydrated. Simply reducing steam and/or water flow rates to the preconditioner may be effective in reducing this buildup. In other cases, the buildup may occur if the tips of the beaters are worn to the point where there is excessive clearance between the beaters and the wall of the preconditioner.

181 Why do I get hard pieces coming out of the preconditioner which later plug the extruder die?

The formation of hard product pieces or buildup on beaters that can later break loose is dependent on the recipe characteristics, operating variables and preconditioner design. Some recipes are very sticky when heated and hydrated and will result in significant buildup on beaters or on the side walls of the preconditioner. If this material is allowed to cool or partially dry, the material will harden and some of the hardened material may actually break away from the metal parts of the preconditioner. These hard pieces may be most evident when the preconditioner is put back into operation. This phenomenon can be reduced by cleaning out the inside of the preconditioner if it is not in operation for more than 6 hours. If the preconditioner is not cleaned, it will be necessary to observe the first 20 minutes of operation carefully during subsequent runs as this is when the larger, hard pieces may be released from the inside of the preconditioner. This may not be a concern if the die orifices are larger than 6 mm. Smaller openings will be more susceptible to plugging from the hard pieces discharging from the preconditioner.

182 Why does steam blow out from the preconditioner vent?

Preconditioners that operate at or near maximum capacity need to be vented (see Fig. 3). The purpose of the vent is to release excess steam and moisture and to vent air that is forced out of the extruder barrel as the product is compressed. The steam injection nozzles that are located on the preconditioner are designed to inject the most steam in such a way that it condenses into the product. To maximize steam

Fig. 3 Excess steam and product particulates escaping from the vent of a preconditioner.

absorption requires a manageable steam velocity to be maintained. Increasing the number of steam injectors or the size of the steam injection orifices will help reduce velocities. Maintaining proper pressure of the steam supply is critical. Recommended steam supply pressures are 30–35 psig (2–2.5 bar). Higher pressures usually result in excess steam that is not absorbed escaping from the preconditioner vent. In addition, the usual temperature of most recipes only allows a product to have an absorbance capacity of around 8% steam. Adding more steam than this into the preconditioner results in excess steam escaping from the vent

183 Why does product blow out from the preconditioner vent?

The same vent described in Question 182 can also be source of some small product particulates discharge. The discussion that centered on reduction of steam from this vent is also applicable to the reduction of product particulates. Recipes that contain very light density components such as fiber sources will tend to have more particulates discharging from this vent. The addition of more water in the preconditioner can reduce this discharge.

184 Is it ever acceptable to close the vent on a preconditioner?

It is not acceptable to close the vent on a preconditioner unless that preconditioner is operated at 50–60% of its rated capacity and steam addition is limited to only

Fig. 4 Waste recycling system module shown next to a single screw system.

about 50% of the normal rate. A better alternative for eliminating the problem of excess steam and product particulates discharging through the vent is to attach a waste recycling system (WRS) to the vent that will capture and contain these exhausts.

185 What are the features of a typical waste recycling system (WRS)?

A WRS is a reclamation system to recover wet under-processed product from both the preconditioner and the extruder that cannot be recycled through the dryer as dry re-work.

Features of such a system include the following (see Fig. 4):

- A heavy duty stainless steel tank with variable-speed impeller for liquefying wet under-processed material.
- A positive displacement pump and mass flow meter for returning slurry to the preconditioner.
- An optional preconditioner vent collection system for capturing excess steam and product fines from the preconditioner vent.
- An optional vacuum draft-assist on the preconditioner vent.
- A control package to operate the system manually (with optional automated package).
- An optional lift system to load tank from collection vessels.

The benefits of a WRS include the following:

- Wet, under-processed product from extruder/preconditioner startup/shutdown is reclaimed and added back into the process, eliminating the need for off-premises disposal in landfills, livestock feeders, etc.

- Improved housekeeping in extrusion area by collection of product fines discharging from the preconditioner vent.
- Improved operator comfort by reduction of steam and product fines discharging from the preconditioner vent.
- Extended life of preconditioner bearings and seals as excess steam is removed in the preconditioner.
- Generation of hot water for the liquefying step, process water, or other unit operations.

The value of a WRS includes the following:

- Housekeeping and sanitation costs are reduced by up to 20% in the extrusion bay.
- Certain preconditioner maintenance and labor costs are reduced by up to 90%.
- Under-processed product disposal costs are reduced by up to 80%.

The specifications of a waste recycling system could include the following:

- The system is not designed for and is not meant to be used to recycle dry fines. Fines generated in the system that are dry (moisture less than 13%) are best recycled by grinding and adding back to the raw recipe batch mixer at rates of 0–20% (maintaining a consistent level).
- The system is meant to collect 'wet' fines and under-processed product from the startup of the extruder and from the preconditioner vent. The typical amount of under-processed product generated equals 1% of the extrusion rate each hour (for example, if the rate is 9000 kg hr^{-1}, the system is sized to recycle 90 kg hr^{-1}).
- The system works by employees manually loading under-processed product into the tank or by collecting it in wheeled carts and loading it into the tank using a hydraulic lift mechanism. A tank capacity of 730 liters (200 gallons) of water will accept a load of 90 kg under-processed material resulting in a slurry of about 12.5% solids. This slurry is pumped back into the preconditioner at rate of 10% of the dry feed rate resulting in 10% of 12.5% (or 1.25%) of under-processed product being continually added to the preconditioner. As the system runs, the slurry becomes more dilute and will eventually become 100% process water again if no more material is added to the tank. The level of solids is low enough, even in the most concentrated situation, for this gradual dilution not to affect the process.
- If a plant generates more than 1% under-processed product (during an extended startup or several startups in less than an hour), the material must be fed back gradually into the tank, not exceeding the 1% rate each hour.
- The tank has a makeup water supply that keeps the water at the desired level in the tank. The makeup water can come from the extruder barrel cooling water or any source in the plant where potable water might be normally 'wasted' or dumped down the drain.
- The slurry tank contains a high speed impeller capable of dispersing wet product discharging from a preconditioner or final extruder die. The impeller typically runs at high speeds for five minutes after each product addition and can then

rotate at slower speeds as required to maintain suspension. The impeller can also be manually operated as needed.

- The slurry from the tank is added as liquid slurry to the preconditioner by a pump and mass flow meter and can be 'slaved' to the dry feed rate.
- The tank size is purposely kept small so that the contents are emptied or exchanged at least once per hour. This prevents accumulation of large quantities of slurry with the associated odor and microbial problems.

186 Why can't I achieve higher throughputs when I close the vent on the preconditioner?

The extrusion process compacts or densifies material as it passes through the extruder. As a result of this deaeration, some air is 'squeezed' back out of the extruder inlet and can travel back through the preconditioner to the vent opening. In addition, excess steam that is not absorbed in the preconditioner (and extruder) will 'pressurize' the preconditioner if it is not vented (see Fig. 5). Closing the vent is acceptable and will not usually cause process problems if the preconditioner is operated at 50% capacity and steam is added at 4% of the feed rate or less. If the vent is closed at higher throughputs or during higher usages of steam, the preconditioner will become pressurized and will be likely to plug within 5–10 minutes of operation requiring that the system be shut down. The pressurized preconditioner can also affect the function of the loss-in-weight feed system metering feed into the preconditioner.

Fig. 5 Pressurized preconditioner resulting from a closed vent.

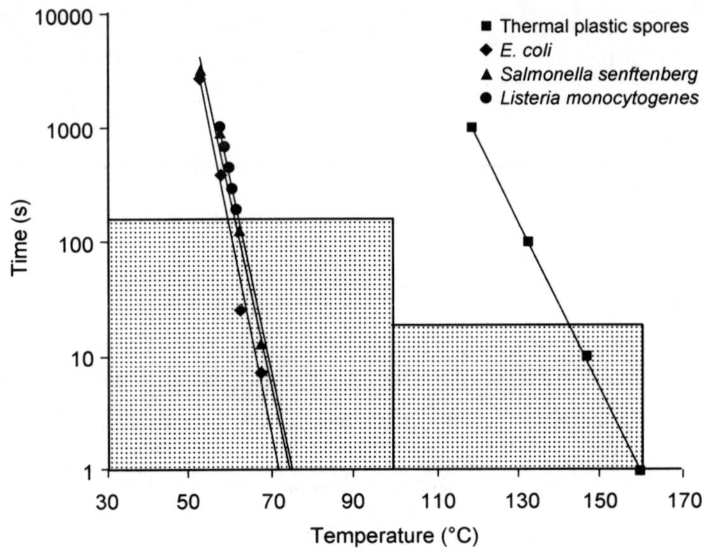

Fig. 6 Thermal death time (TDT) curve for common microorganisms.

187 How can I prevent the occurrence of salmonella in the material coming out of the preconditioner?

One of the primary reasons for utilizing a preconditioner is to heat and hydrate material to the point that the starchy components begin to gelatinize. Another primary reason for the heating step is to help control or reduce salmonella levels in the material being processed. Salmonella is a food safety concern in many products but this microbe is susceptible to fairly moderate temperatures. It is generally accepted in the industry that if temperatures of 77 °C or higher are attained, salmonella will be destroyed (see Fig. 6). These temperatures can easily be attained during the preconditioning step of most extrusion applications, making the preconditioning stage a very effective control point for salmonella destruction.

188 What is the retention time distribution and how is it measured?

The retention time distribution (see Fig. 7) is an indication of the relative variation in retention times of all particles within the preconditioner. It is an indication of plug flow or how well a preconditioner maintains a 'first in, first out' flow of material through the system.

All particles that pass through the preconditioner should ideally have exactly the same retention time and the resulting curve in Fig. 6 would actually be a vertical line at that retention time. In reality, the curve takes on the appearance of the curve generated in Fig. 6 indicating that most, but not all, of the particles have

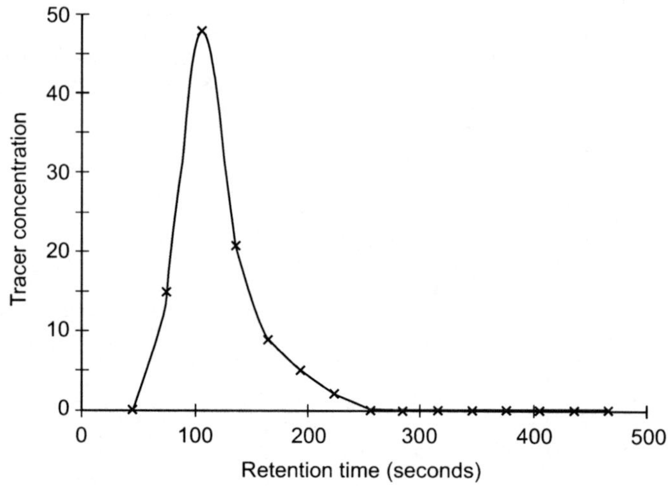

Fig. 7 Example retention time distribution curve for a specific preconditioner.

a similar retention time. The horizontal axis is in units of time while the vertical axis is units of particulate concentration. The retention time distribution curve can be established by introducing a tracer into the inlet of the preconditioner during operation and then measuring the concentration of that tracer as it discharges from the preconditioner over a specific time interval.

189 How can I measure retention time in a preconditioner?

Measurement of retention time is another useful indicator of preconditioner performance. This measurement should not be confused with retention time distribution as discussed in Question 188. Retention time is really an indication of the average retention time, that is, the amount of time for which material remains in the preconditioner during operation. Retention time can vary with each set of operating variables and can be affected by product density. The easiest method of measuring the average retention for a particular product and a specific set of operating parameters is to determine the amount (mass or weight) of material within the preconditioner at any given moment. The preconditioner is operated with the normal feed and water/steam addition rates employed during extrusion as if actual product were being processed. The rate of material discharge of the preconditioner is determined by weighing the flow over a given period of time (for example, 30 s). This measure determines the flow rate through the preconditioner. After this measurement is complete, the entire system (including feed delivery, preconditioner rotation and steam and water addition and discharge) is stopped. The material remaining inside the preconditioner is removed and weighed. This weight, divided by the rate previously determined, will yield the average retention time.

190 Why do I get different product temperature readings even when I have the same process conditions?

Product temperature readings are usually made by inserting a thermocouple or similar temperature sensing device into the product flow stream as it discharges from the preconditioner. Accurate temperature readings may be hindered by several factors. If product buildup occurs around the tip of the thermocouple, the product can serve as an insulator and the temperature readings may be altered. The amount of buildup is dependent on the location of the sensor, the recipe characteristics and the amount of steam and water that are added during the preconditioning step. The sensor location is critical. If it is located in a position where the product flowing through the preconditioner does not fully cover the exposed tip, excess steam from either the preconditioner injection site or from the extruder barrel can give readings on the high side.

191 How do I prevent wear to my preconditioner components?

Wear is a natural progression or phenomenon that will occur with any physical apparatus as a result of normal operation. Abrasive wear is the most common type experienced in a preconditioner and is the result of recipe material contacting the internal components of a preconditioner as it is being conveyed through the preconditioning chamber. The normal life of preconditioner components as a result of this type of wear is 7–10 years or longer. When this age range is reached, the tips of the beaters may be worn or rounded to the point that the clearance between the tip of the beater and the inner wall of the preconditioner chamber is great enough to significantly broaden the retention time distribution and allow product build up on the walls.

Extended wear metallurgy could be used to lengthen the life of these parts but may not be cost-effective. Operating the system at higher moisture levels through increased steam or water addition would also extend the wear life. Obviously the recipe components have a big impact on the rate of wear. More abrasive ingredients such as rice bran and certain minerals will increase the rate of wear.

192 Why does my preconditioner experience corrosion?

Corrosion is another type of wear that is sometimes experienced in preconditioners. Usually corrosion is a concern when the pH of the recipe substrate is less than 5 or high levels of various salts are present. The two approaches to reducing this type of wear are to remove the corrosive material from the recipe or to construct product contact points inside the preconditioner with materials that are corrosion resistant.

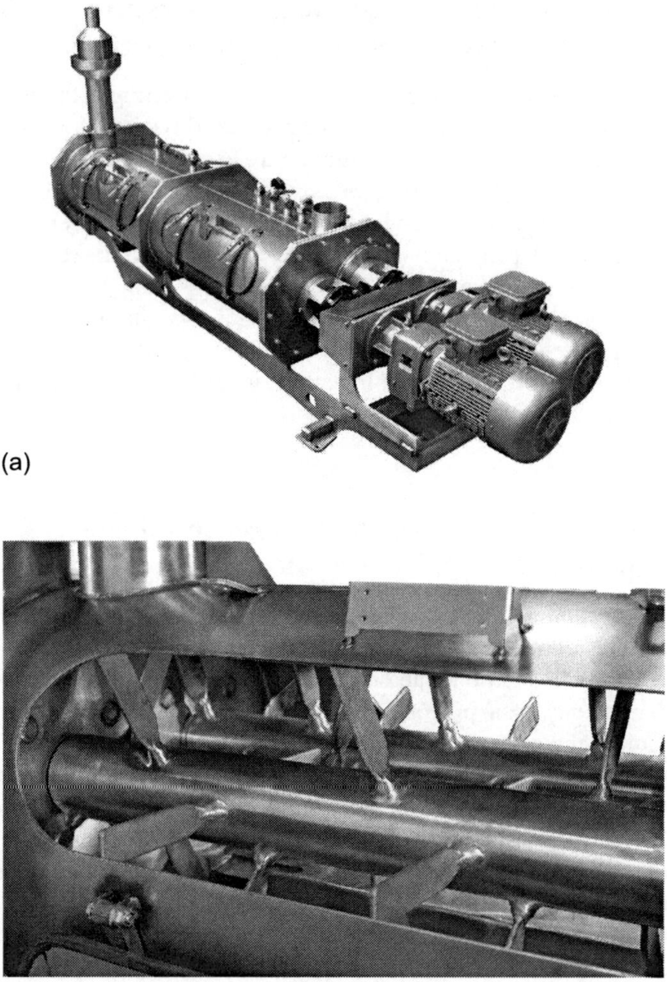

(a)

(b)

Fig. 8 (a) Double-shafted preconditioner, (b) beaters in a double-shafted preconditioner.

193 What is a double-shafted preconditioner?

A double-shafted preconditioner (see Fig. 8(a) and (b)) is a chamber containing two parallel shafts that each contain beaters rotating in a desired direction and speed. Mixing intensity is much greater with a double-shafted design compared to a single-shafted design. Most modern extruders today utilize the double-shafted design as it enhances the extrusion process and reduces wear on extruder components.

194 Are vertical preconditioners available and when would this design be useful?

The vertical preconditioner is an older design that used to be utilized in those applications in which a minimum amount of preconditioning was required. The major disadvantage of this design is that there is almost no assurance of plug-flow or first in, first out. In the vertical design, raw material was metered into the top of the preconditioner and then cascaded down through it, exiting at the bottom. Mixing intensity was difficult to control in this vertical design. Because of these process challenges, the vertical design is seldom used in food extrusion applications.

195 What design parameters should I consider when selecting a preconditioner?

Preconditioners are evaluated by comparing their performance in terms of measured starch gelatinization, retention time, mixing efficiency and residence time distribution. Preconditioner designs are evaluated by comparing sanitary features and accessibility. The objectives of preconditioning are to heat, hydrate and mix the recipe. The efficiency of a preconditioner is how well the following objectives are accomplished:

- Heating: energy inputs equal useful energy out plus losses.
- Hydration: uniformity.
- Mixing: coefficient of variation.

196 How does the paddle configuration affect mixing in the preconditioner?

The paddle configuration affects product retention time. The number of paddles and the paddle tip speed affect mixing. Increased retention time results in increased mixing. A larger number of paddles and a greater paddle tip speed increase mixing.

197 When is it not acceptable or necessary to have a preconditioning stage?

The use of a preconditioning stage in the process is warranted if process moisture exceeds 18%. If the recipe contains large particle sizes that are difficult to hydrate and cook, preconditioning is indicated. If these two conditions are not present, the preconditioning step may not be required.

198 How do I predict or determine what conditions are required to precondition a new product?

Steam addition is usually involved in preconditioning a product. Theoretically, it is not possible for most recipes to absorb much more than 8% steam addition (% of the dry recipe feed rate) if the starting raw material is at room temperature. Water in the preconditioner is added at levels necessary to attain the correct process moisture. In an ideal process, about one-third or more of the cooking (measured in starch gelatinization) should occur in the preconditioning phase

Single screw extruders in food and feed extrusion: common problems and their solutions

199 Why does product back up into the inlet of the extruder?

Several factors can contribute to product backing up into the inlet. All of these items are a factor because they reduce the conveying or transport power of the extruder barrel:

Feed rate exceeding the volumetric capacity of the extruder screw
If feed rates are attempted that exceed the volumetric capacity of the extruder screw volume, the product will eventually back up into the extruder inlet. The volumetric capacity can easily be increased for a given extruder bore by simply increasing the screw speed. At some point screw speed may become too high for the design of the various extruder drive components. The extruder manufacturer should always be consulted before changing the screw speed range. Volumetric capacity can also be increased by installing a larger bore (larger screw diameter) extruder barrel.

Excess steam injection
Normal steam injection rates (into the extruder barrel) are in the 1–4% range (% of the dry feed rate). Injection rates above this range may prevent feed from being transported past the steam injection zone causing the product to back up into the inlet. Simply reducing the steam injection levels will remedy this problem.

Excessive restriction within the extrusion process
Partially blocked extruders (from mechanical devices such as internal locks or

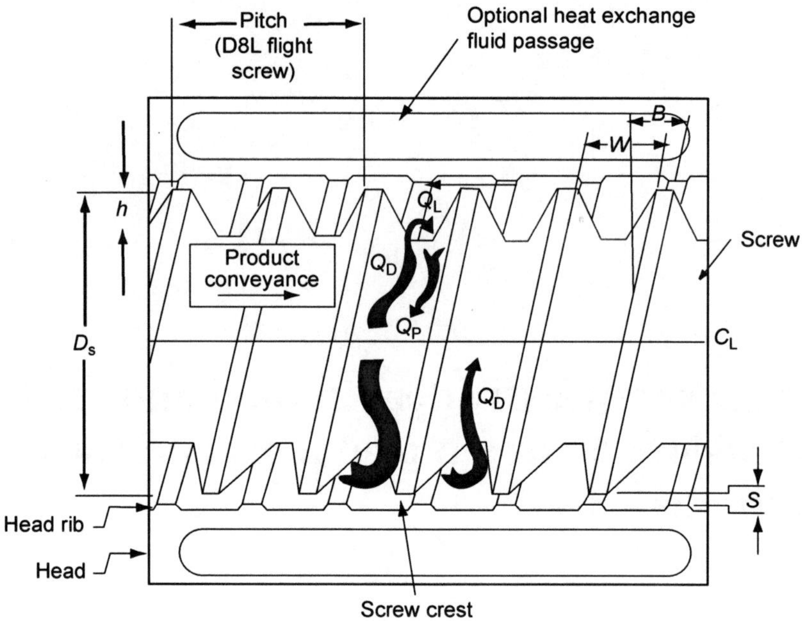

Fig. 1 Product flow patterns inside a single screw extruder. *W* is the width between screw flighting, *B* is the angle of screw flighting with respect to perpendicular of center line, C_L is the center line of the screw/extruder barrel, *h* is the height if screw flighting, D_S is the diameter of screw, *S* is the gap between outside diameter of screw and insider diameter of extruder barrel liner wall, Q_D is the drag flow coefficient, Q_P is the pressure flow coefficient and Q_L is the leakage flow coefficient.

final assemblies) can create an overfill in the extruder barrel. Examining the barrel setup and reviewing the amount of open area in the die are also useful to ensure that these are not too restrictive.

Worn screw elements and barrel liners

The transporting power of screw elements is severely compromised when the elements become worn. The leading edge of the screw flighting becomes rounded increasing leakage (Q_L) flow (see Fig. 1), which in turn increases barrel fill and eventually causes product to back up into the inlet. Worn or partially worn barrel liners can also decrease the ability of the screws to transport material through the extruder and could eventually result in a plugged inlet section.

200 Why does the extruder overload?

If it is assumed that the term ‘overload’ means that the main drive motor is pulling full load or more than full load, this overload condition can be a temporary spike or a sustained position in the red zone. A temporary spike into the overload zone may be the result of several factors:

- A sudden 'slug' of feed fed into the extruder: this may come from non-uniform metering of feed into the system or, more likely, a buildup of feed that suddenly frees itself and is fed into the barrel.
- When product builds up or temporarily sticks to the screw or preconditioner surfaces and then entering the product stream.
- Sudden and intermittent loss of water flow to the preconditioner or extruder barrel.

A sustained overload condition can also be the result of one or more factors:

- A feed rate setting that is too high for the process: usually reducing the feed rate until the motor load is within a safe operating range will correct this problem.
- If the extruder configuration is too aggressive, too much power is consumed resulting in an overload. Provided the product technical qualities will still be met, modifying the extruder configuration to produce more transport and less friction and barrel fill will correct this condition.
- Too much die restriction: too little die open area, too long land length and too much distance from the screw end to the die entrance can all require more power. There could be a partial or complete blockage of several die openings which would increase barrel fill and result in an overloaded condition.
- Recipe characteristics are such that viscosities remain high even when heated or subjected to shear; some recipe components can actually become sticky and very elastic during extrusion which can result in excessive power requirements.

The addition of a lubricant such as oil or fat at low levels into the recipe can reduce viscosity. Increasing extruder steam injection levels into the product can reduce viscosity and decrease load. The addition of water to the preconditioner or extruder can reduce viscosity.

Excessive wear on screw components can result in abnormally high barrel fill. This condition is discussed in detail in Question 199.

201 What is causing the product to surge – a change in size and shape coming from the extruder?

Several factors are responsible for the condition known as 'surging' (see Fig. 2):

- If the level of steam injection into the extruder barrel is too great, it is likely that product surging could result. It is best to spread the steam injection quantity over several injectors rather than concentrate the injection in one injector.
- High extruder barrel temperatures are often responsible for product surging. Simply decreasing product temperature set points to initiate cooling will decrease surging.
- It may be helpful to increase the level of water addition to the extruder barrel to decrease product viscosity and stabilize the process.
- If the steam source is not properly plumbed to remove condensate, surging may be the result of intermittent presence of condensate in the process steam supply.

Fig. 2 Variations in product size caused by surging.

- Check to make sure the preconditioner vent is not plugged as this may pressurize the preconditioner which could interrupt material flow and result in surging.
- Excessive screw or sleeve (liner) wear can result in unstable transport of product through the extruder resulting in product surging.
- If the open area in the die is too great, product surging may be evident.

202 What is the best method of controlling product temperature?

Product temperature can be controlled in a number of ways. Product temperature is the result of the amount of energy transferred into or out of the process and any exothermic or endothermic chemical reactions that may be occurring. Energy inputs are the result of mechanical or thermal energy applied to the process. Mechanical energy is measured by the power consumed by the extruder main drive motor. Higher loads mean more mechanical energy input and thus higher product temperatures. Thermal energy inputs are mostly from steam or water injection into the preconditioner or extruder barrel. More steam inputs and more and hotter water inputs increase product temperature. Product temperature is also influenced to a lesser degree by heating or cooling mediums circulated through the jackets of the extruder barrel. This process must depend on the influence of heat transfer through the barrel liners and is not as influential as direct steam or water injection. Control of product temperature is the result of all of these process variables.

203 Why do product moisture and expansion levels vary when adding steam into the extruder barrel?

Direct steam injection into the product inside the extruder barrel is a very efficient method of getting energy (thermal energy) into the process. As steam is added, the increased energy results in greater product expansion at the die. Steam is also a source of moisture. One kilogram of steam is equal to one kilogram of moisture into the process.

204 How do we prevent product from over-expanding from the die?

Products that are over-expanded result in final bulk densities that are too low and create problems in packaging with over-filled bags that are difficult to close properly. Excessive expansion is generally the result of excessive energy input into the extrusion process. Specific factors that reduce over-expansion are as follows:

- Decrease the direct process steam injection into the product in the extruder barrel.
- Decrease the process water level to the preconditioner as required.
- Reduce product temperature in the preconditioner by decreasing the level of direct steam injection into the product.
- Reduce the product temperature in the extruder barrel by lowering the temperature setpoints along extruder barrel, resulting in more cooling water circulating in the head jackets.
- Decrease the level of addition of extruder barrel process water into the product.
- Reduce the dry feed rate.
- Increase the open area of the die.

205 What causes my products to be constantly distorted in shape?

Extruded products can become distorted in a number of ways:

- Certain recipes yield a very tender and soft product which can become distorted when cut at the die or when conveyed to downstream unit operations such as drying or cooling.
- Knife blades require a minimum distance between the actual cutting edge of the blade and the die surface. If there is a large gap at this point, the product cutting will be compromised and result in a distorted product.
- If the product is conveyed from the extruder die to the dryer or cooler via a negative pneumatic conveying system, the air velocity needs to be adjusted to minimize product distortion or damage.
- Partial plugging of a die orifice will consistently distort the product.
- If products are conveyed by mechanical conveyors after the extruder, the product is often contained in the die/knife area by a product hood which, if too restrictive, can distort product that impacts the inside of the hood or shield at high velocity.

206 How can I increase cook in the extruded product without over-expansion?

Usually an increase in cook or energy input results in a more highly expanded product. There are several methods that can be used to produce a highly cooked product that does not have excessive expansion.

- The extruder barrel can be divided into a primary cooking zone (to cook the product) and a secondary zone to cool and densify the product to prevent expansion.
- Another process method that is employed is to couple a cooking extruder with a second forming extruder where the fully cooked product is cooled and densified to avoid expansion.

207 What causes a burnt appearance and taste in extruded products?

A toasted or scorched appearance and taste in products can vary from something that is almost undetectable to obvious discoloration and a strong burnt flavor (Fig. 3). The attributes in the technical quality of a product are either from processing at too low moisture levels or too high extrusion temperatures or a combination of the

Fig. 3 Sample of scorched product.

two conditions. The simple correction for this condition usually requires an adjustment in product moisture during extrusion by increasing water addition. A second and more likely factor that contributes to scorched product is over-drying product at excessive temperatures in the drying operation. Reducing drying temperatures and/or dryer retention times will also help correct this phenomenon.

208 How can I maintain proper product density?

This question was partially addressed in the answers to Questions 204 and 206 earlier. Product density is a result of many factors including recipe characteristics, processing conditions and hardware components. Assuming recipe constituents and hardware components are constant, the following operating parameters are varied to control product density:

Mechanical energy input

- adjusting screw speed
- adjusting die restriction
- adding internal devices to the extruder which can vary restriction while the system is in operation.

Thermal energy input

- adjusting process steam to preconditioner and extruder
- adjusting process water to preconditioner and extruder
- adjusting extruder barrel temperatures to influence heat transfer.

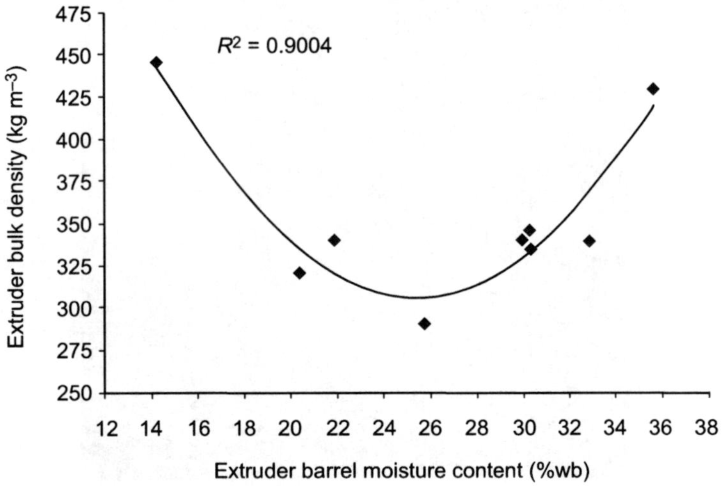

Fig. 4 Effect of extrusion moisture on product bulk density.

Some effort has been made in the industry to develop control systems where the product density is measured on line and this particular product technical quality is then controlled automatically by adjusting steam levels, venting procedures and using variable restrictive devices.

The most common practice for controlling product density is to measure bulk density of the product coming off the extruder manually and then adjust extrusion parameters such as steam or water input to bring the density into specification. The correlation between extrusion moisture and product bulk density is shown in Fig. 4. Note that as the moisture of extrusion is increased, the product density first decreases and then, after reaching a minimum density, will gradually increase again as moisture continues to increase.

209 What causes the extruder to leak along the length of the barrel?

Occasionally an extruder will have product leaking out of the joints between the barrel segments as indicated in Fig. 5. Although the extrusion sections are designed to withstand high operating pressures, an occasional sharp spike or increase in

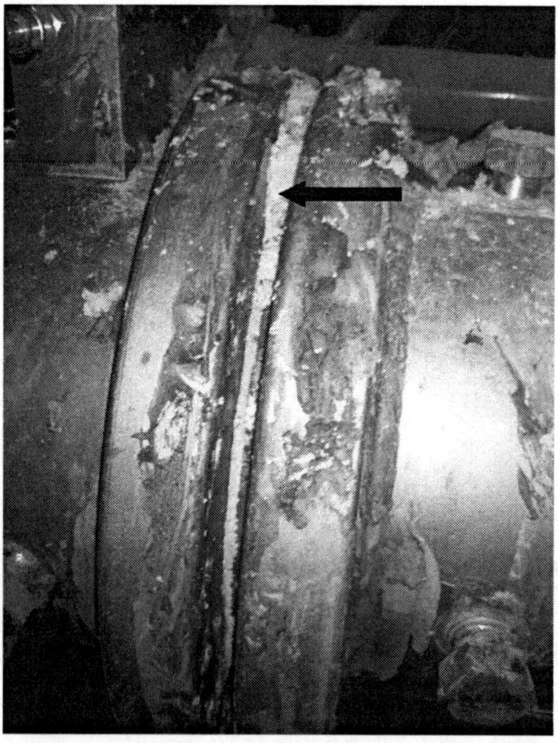

Fig. 5 Product leakage between head sections.

product pressure can momentarily result in some product leakage. This is not common and is usually the result of two factors:

1. The mating surfaces between the head sections have been damaged or marred by a sharp instrument and will not seal tightly due to raised areas or depressions.
2. When the head sections have been reassembled after separation for inspection, cleaning, or extruder configuration changes, a failure to make sure that all residual product or other material was removed from the mating surfaces will result in minute gaps which lead to product leakage during subsequent operation.

If either items 1 or 2, have occurred, it is likely that the bolts securing the head sections together may have stretched slightly in length and thus may not have been fully retightened resulting in gaps between the heads causing additional product leakage. The proper grade of hardened bolts must be used for the application.

210 Why do my process steam and water injection levels vary during extrusion?

Unstable flow rates of any of the mass inputs to an extrusion process can be disruptive for uniform product size, shape and textural properties. It is particularly important for water and steam flow rates to be stable. If these flow rates vary too far on the low side, it is possible that the extruder could lock-up owing to dry feed seizing the screw inside the barrel. There could be several reasons why variations in flow rates occur:

- Pressure variation in the steam and water supply lines. Usually a pressure regulator is inserted in the plumbing harness to avoid or reduce this variation.
- Automated flow control valves may experience 'chatter' or other interference that results in flow variation.
- The product fill in the extruder barrel may vary and occasionally plug or partially plug the injection sites of process water and/or steam to the extruder barrel. As this fill goes down or is reduced, the steam and water flows will transition back to normal flow rates.
- As product fill in the barrel increases, steam and/or water flow to the extruder may be partially or momentarily plugged, thus affecting the flow rates.

211 How can I read a true product temperature inside the extruder?

Knowledge of the true product temperature inside the extruder is important for many reasons, but a primary reason is to help ensure food safety by monitoring process temperatures required for pasteurization of the product. True product

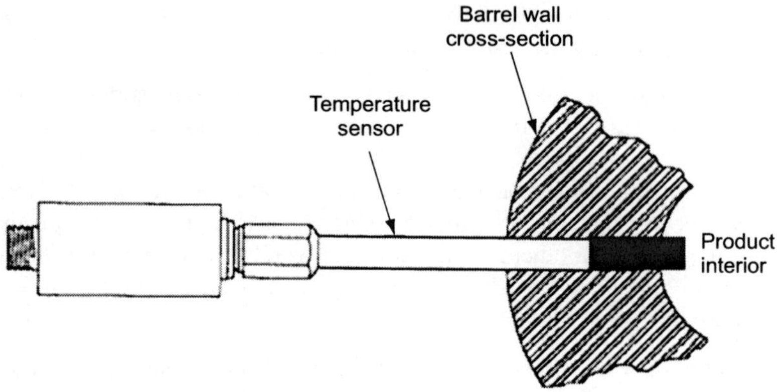

Fig. 6 Location of temperature sensor.

temperatures are best measured in an extruder by placing a temperature sensor near the end of the extruder after the screw. Several key points to remember when choosing how and where to mount the temperature sensor are as follows:

- Mount the temperature sensor after the screw and near the end of the extruder where there are no jacketed surfaces where the heating or cooling medium would adversely affect true product temperature readings. Placing the sensor in a die spacer or adapter between the end of the screw and the final die is an ideal location.
- The tip of the sensor should be inserted at least 6 mm into the product stream from the side wall of the extruder (see Fig. 6).
- The sensor tip is often protected from excessive wear by a metal sheath.
- It is recommended that where the temperature sensor passes through the side wall of the extruder it should be protected from convectional influences by an insulated sheath.

212 What causes the extruder barrel to move in an eccentric fashion?

The extruder barrel is normally supported in some way by the frame on which the whole system is mounted. If the extruder barrel moves, and this can happen sometimes even when the barrel is supported, it can indicate that the extruder shaft is bent, the barrel is misaligned, or a combination of extruder configuration and recipe characteristics result in a product mass that is not fluid or flowable inside the barrel.

Bent shafts will cause the end of the extruder barrel to move in an eccentric fashion and this can be corrected by straightening the shaft or, in some severe cases, replacing the shaft. Proper alignment of the extruder barrel sections around the shaft will also reduce this movement. Barrel alignment techniques should be taught by a qualified service engineer experienced in the maintenance of the system.

Some recipes do not lose viscosity during the heating/hydrating processing steps normally present in a cooking extrusion environment. This fact, coupled with a configuration that utilizes a screw profile that may become unbalanced when extruding at certain operating parameters, can result in the extruder barrel moving in an eccentric motion.

If operation of an extruder that has severe eccentric movement in the barrel continues, fatigue may result in the shaft and barrel components which may lead to costly repairs and unscheduled down time.

213 How can I eliminate the knocking noise the extruder makes when it is in operation?

The extrusion service engineer will occasionally find a situation where there is a 'knocking noise' coming from the extruder barrel. Although this is not common and not to be expected, it does not necessarily indicate a critical situation requiring emergency action. The severity and loudness of the knocking determines the seriousness of the situation. Some extruders that are slightly out of alignment will emit a knocking noise during the startup phase until product fills the extruder barrel and serves as a bearing to prevent the screw/barrel liner contact which is responsible for the noise.

To reduce or eliminate this noise requires the following actions:

- Check the extruder shaft to ensure that it is not bent and straighten it if necessary.
- Align the extruder barrel around the shaft as required.
- Install a shaft-centering device (a stationary shearlock) which prevents screw/barrel liner contact and can reduce wear by up to 30%.

214 What causes extruder throughput to decrease over time?

Some of the same factors discussed in Questions 199 and 200 are likely to cause decreased extruder throughput over time. Wear on the screw elements and barrel liners will gradually increase extruder load and require throughputs to be reduced to prevent an overloaded condition.

215 The wear on my extruder components is excessive – what can be done about this?

It is important to define what is meant by the term 'excessive'. Products vary in their sensitivity to worn extruder components. In some processes, it may be

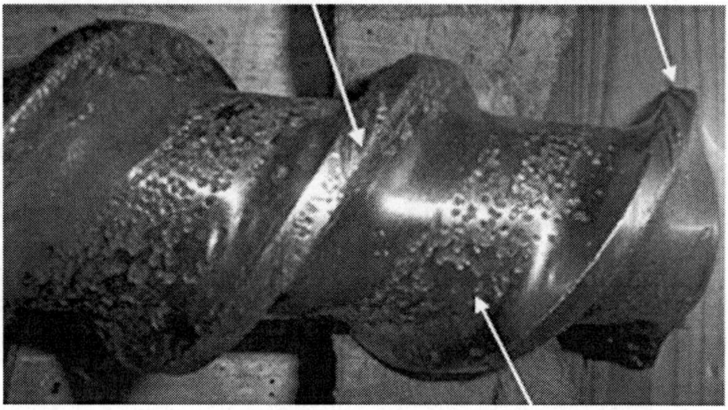

Fig. 7 Examples of common wear patterns.

possible to extrude acceptable product successfully even with significant wear. In general, the process can be expected to become unstable due to wear that results when the gap between the tip of the screw element and the inside diameter of the barrel liner is 2.5 times greater than new or unworn conditions. Wear is the result of numerous factors (see Fig. 7). Wear patterns normally develop from abrasive action (this is normal wear from friction between product and screw and liner surfaces).

Decreasing the particle size of the recipe components can be an effective way to decrease wear. Smaller particles cook more easily and are less abrasive. Increasing moisture content during extrusion can reduce wear as this generally results in lower mechanical energy input which is a variable that is directly related to abrasive wear. If the recipe allows flexibility, the addition of fat or oil may reduce friction as these ingredients act as a lubricant. A review of the screw and liner metallurgy is necessary as this component can be adjusted through selection of metallurgy that is more abrasive resistant.

Corrosive wear is usually the result of low pH (acidic) conditions, excessive use of salts in the recipe, or screw and liner components not manufactured for the proper metallurgy requirement for the process environment within the extruder barrel.

Another wear pattern that is occasionally seen is called forging wear. Forging wear is the result of metal-to-metal contact and can often be accompanied by an audible knocking noise during operation of the extruder. Forging wear is usually the result of a bent shaft, misaligned barrel, or, in some cases, improper support of the extruder shaft. Realigning the barrel and making sure that the shaft is straight are the first steps to correct this situation. There are mechanical devices such as internal bearings in the extruder barrels that are used to support the shaft and keep it centered within the barrel. The use of these devices can decrease barrel component wear by as much as 30% or more.

Fig. 8 Extruder design where barrel components are held together by three rods.

216 How should an extruder be safely restarted after a power outage?

Power outages can cause unexpected shut downs of an extrusion system. If the power is off momentarily (less than 1 minute), most systems can be restarted immediately. Although this ability to restart the system immediately can vary with recipe characteristics, most systems will not cool enough to allow the product viscosity to increase significantly.

If the power is off for a longer time, it is best to remove the die assembly immediately and attempt to restart the extruder motor. If the motor will not rotate the screw when the die is removed, it is advisable to remove the cone head as quickly as possible. Removal of the die and cone head eliminates the greatest restrictions, depending on barrel fill. After removing the cone head, an attempt should be made to restart the system again. Never attempt to restart an extruder when the individual heads are not stabilized. In some extruder designs (Fig. 8), all of the heads are held in place by two or three support shafts capped with nuts. Once these nuts are removed and an attempt is made to rotate the screw, it is possible that the complete barrel will rotate with the screw. This is not only dangerous but can damage extruder parts and plumbing systems.

Where extruder heads can be removed individually and the remaining heads are still held in place by some means, it is safe to continue to remove one head at a time always followed by an attempt to rotate the shaft and screws. This procedure should be repeated until the shaft and screws rotate freely and the product plug is removed or conveyed out of the barrel.

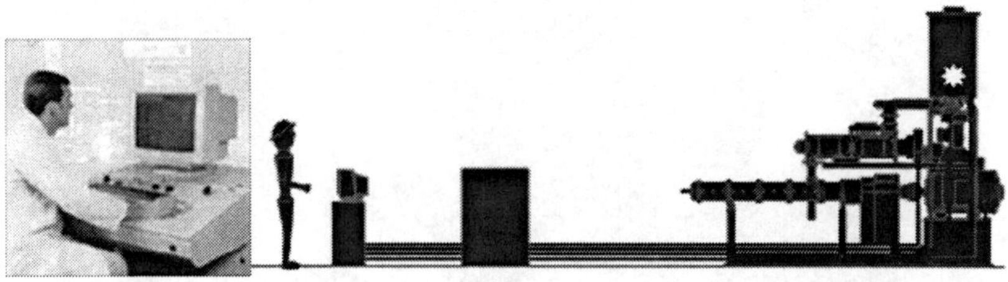

Fig. 9 Modern computer control systems.

217 How can I increase the range of ingredients processed on the single screw extruder?

Single screw extruder systems have been the workhorse of most of the industries that utilize extruders for many years. In these industries, the range of products produced is increasing annually. As the spectrum of products produced broadens, more and more flexibility is required of extrusion systems. In most industries that utilize extrusion, the single screw extruder can produce 90–95% of the required products. The other 5–10% of products require twin screw technology because of its self-wiping fully intermeshing properties which result in a more positive pump that can handle sticky ingredients which would potentially build up in the root diameter of the screw elements.

The processing range of single screw extruders has gradually increased over the years owing to the following technological advancements:

- **Variable speed drives**: Twin screw extruder systems have always enjoyed variable speed drives and adopting this technology for the single screw extruder has greatly broadened the flexibility of the single screw extruder.
- **Computer control systems**: The use of automated control systems (Fig. 9) for mass flow and temperature control has improved the stability and control of the single screw system.
- **Screw configuration profiles**: New profiles for the screw elements in single screws has improved both conveying and mixing or kneading properties as required for various properties.
- **Die technology**: Streamlined die flows and on line, variable restrictive devices have permitted more control of the single screw extruder process.

218 Who can help me select the proper extruder configuration?

When a new extrusion system is purchased, the vendor (extruder manufacturer) should supply the extruder with an appropriate configuration for the application

Fig. 10 Consulting services are available for extrusion hardware selection and process development.

(Fig. 10). This configuration may have to be modified during or after commissioning owing to ingredient variations or discrepancies in the final product properties. The extruder manufacturer should be contacted when a used system is purchased. They would also be able to give advice and provide the necessary hardware components for updating older technology. Most extruder manufacturers maintain a technical center where test runs can be conducted on smaller laboratory extruders to confirm the proper selection of hardware before purchases are made.

There are a number of consulting enterprises with experienced associates that are available to assist in selecting extrusion hardware and developing a process. Experienced personnel within your organization can also be a source of valuable input.

Twin screw extruders in food and feed extrusion: common problems and their solutions

219 Why does product back up into the inlet of the extruder?

Several factors can contribute to backing up of product into the inlet. All of these items are a factor because they reduce the conveying or transport power even with a twin screw extruder.

Feed rate exceeds the volumetric capacity of the extruder screw
If feed rates that exceed the volumetric capacity of the extruder screw volume are attempted, feed will appear in the inlet and eventually result in a choked or plugged condition. Volumetric capacity can easily be improved for a given extruder bore by simply increasing the screw speed. At some point screw speed may become too high for the design of the various extruder drive components. The extruder remanufacturer should always be consulted before changing the screw speed range. Volumetric capacity can also be increased by installing a larger bore (larger screw diameter) extruder barrel.

Excess steam injection
Normal steam injection rates (into the extruder barrel) are in the 1–4% range (% of the dry feed rate). Injection rates above this range may prevent feed from being transported past the steam injection zone causing the product to back-up into the inlet. Simply reducing the steam injection levels will remedy this problem.

Excessive restriction within the extrusion process
Partially blocked extruders (from mechanical devices such as internal locks or

final assemblies) can create an overfill in the extruder barrel. Examining the barrel setup and reviewing the amount of open area in the die are also useful to ensure that these are not too restrictive.

Worn screw elements and barrel liners

The transporting power of screw elements is severely compromised when the elements become worn. The leading edge of the screw flighting becomes rounded increasing leakage flow which in turn increases barrel fill and eventually causes product to back up into the inlet. Worn or partially worn barrel liners can also decrease the ability of the screws to transport material through the extruder and could eventually result in a plugged inlet section.

220 Why does the extruder overload?

Extruder overload is a descriptive term for a state or condition where the main motor driving the extruder is surpassing full load. Although potential factors that result in an extruder high motor load are listed below, it is possible that the main cause is simply that an under-powered drive motor was supplied for the application. If many of your products are pushing operating parameters to near the full load mark, I would suggest considering installing a larger motor. It is advisable to contact the extruder manufacturer to make sure that larger motors can be installed without exceeding the design limitations of the complete drive system. A guide to trouble-shooting excessive motor loads includes the following checklist:

- **Ensure that the dry feed system is delivering the correct feed rate** and that the feed rate is consistent. Excessive dry feed rates can lead to high motor loads.
- **Lack of adequate process steam injection into the extruder** barrel can also result in high motor loads. Not only must adequate steam be supplied, but the steam flow rate must be consistent. Often the barrel fill will be great enough partially or completely to block the injection ports reducing steam inputs and causing high motor loads.
- **Increasing extrusion moisture** by adding more extruder or preconditioner water will be helpful in reducing motor loads if the process will accept increased moisture levels.
- **Increasing steam addition to the preconditioner** will usually reduce extruder motor load because of the added moisture and added thermal energy, both of which reduce product viscosity and reduce motor load.
- **Insufficient die open area** can result in high motor loads. This may be because the existing die area is decreased because of blocked orifices or because the original open area was not adequate. Every effort should be made to avoid prolonged overload conditions as most electrical drives are protected from extended high amperage loads which can result in overheated electrical systems. Electrical starters or variable frequency drive (VFD) systems will stop

when overloaded for extended times and then the extruder will have to be freed from a plugged condition which is time consuming and frustrating.

221 What is causing the product to surge – change in size and shape coming off the twin screw extruder?

A product surge – change in product shape, size or velocity through the die orifice – is really indicative of extruder instability. This surging phenomenon is also evident in extruder motor load fluctuations. There are a number of factors that could be responsible for this instability in the process.

Inconsistent process steam injection into the product in the extruder barrel
Check to make sure steam flow is not interrupted by inadequate steam supply pressure (recommended to be at least 6 bar) or that the extruder barrel is so full of product that barrel pressures exceed steam pressures resulting in little or no steam flow into the process. At times the barrel fill may actually vary resulting in inconsistent steam flow and product surge.

Extruder head temperature
Excessive head temperatures, especially in the feed sections of the extruder, can cause partial sticking of the product to the side of the barrel and surging in product flow.

Extrusion moisture
Low moisture contents coupled with high extrusion temperatures result in poor flow characteristics within the screw. The simple action of adding more extrusion process water will reduce product viscosity in the barrel and move product above the melt transition temperature into the melt region.

Condensate in steam source
The presence of condensate in the steam lines supplying process steam to the preconditioner and extruder will affect process stability. Condensate changes energy flow inputs when it is present in the steam affecting product expansion and viscosity. Proper installation procedures for steam plumbing will include headers and steam traps to remove or reduce condensate.

Excessive screw wear
As screw elements wear, they become much less efficient in conveying product, especially against any restriction. Although twin screw extruders can be less

sensitive to minor wear compared to single screw systems, they do eventually reach a point where wear is sufficient to affect product conveyance greatly. A general guideline that can be used to judge when wear becomes excessive is to determine the gap between the outside diameter of the screw element and the inside diameter of the barrel liner or wall. When this gap enlarges to 2.5 times the original size when new, wear is likely to affect the stability of the process and is considered to be excessive.

Incorrect die open area

If the total open area of the die openings is too great, some surging may be evident during processing as the extruder is not able to apply equal flow or pressure to all openings. Product will tend to take the path of least resistance and if there are any partially plugged openings or unequal restriction for any reason across the back face (inlet side) of the die, there is likely to be unequal flow or surging through the die.

222 How do we prevent product from over-expansion from the die?

Expansion of product coming from the die is due to the sudden decrease in pressure from the high pressure environment within the extruder barrel to the atmospheric pressure environment of the production floor. Product temperatures within the extruder barrel reach well above 100 °C but any moisture contained in the product remains in the form of water owing to the high pressures. As the product exits the die orifice the sudden drop in pressure allows the moisture to turn to steam and as this steam attempts to escape the product it causes the product to expand, much like an expanding gas within a balloon. The more elastic the product the greater the expansion as the steam is captured for a longer period of time before it escapes into the atmosphere.

Based on the principle described above, many steps can be used to control the expansion and avoid 'over-expansion' which usually occurs when the product is not sufficiently dense. Some of the more common steps employed are as follows:

Reduce product temperature within the extruder barrel

Adding less energy into the process or removing energy from the process are both methods of lowering barrel temperatures and thus reducing product expansion. Many extruder barrels are jacketed or have some means of cooling the barrel by removing energy, often through circulation of cold water or a chilled fluid through these jackets. If steam injection is part of the process variables, reducing the amount of steam used in the process will also lower product temperatures and reduce expansion. This reduction in steam addition can occur in the preconditioner and/or the extruder barrel itself. Any steps to reduce energy input will be effective including reduction of screw speed to reduce mechanical energy inputs. The amount of open area in the die can be increased along with other steps in die design to reduce restriction of product flow through the die assembly with the net effect of less product expansion. Other steps to reduce product expansion include modification of the extruder configuration to move toward a setup with greater transport power and less mechanical energy or work and shear.

Establish a vent within the extruder barrel

Installing a vent in the extruder barrel will drastically reduce product expansion. A vent allows pressure to escape from the extruder barrel and also facilitates evaporative cooling. Both of these steps will affect the process by removing energy from the process and thus reducing expansion. On many occasions a vented opening in the extruder barrel will be fitted with a valve or mechanism to control the amount of pressure release and this becomes a method of controlling expansion.

Extrude product into an enclosed chamber to reduce pressure drop at the die

Air-tight chambers have been designed and manufactured to enclose the die/knife mechanisms completely. The chambers are fitted with a rotary airlock at the bottom to allow discharge of product without losing significant pressure inside the chamber. By controlling the pressure inside the chamber, it is possible to control the temperature at which moisture in the extrudate will boil or turn to steam. Various chamber pressures impact the boiling point of water and thus impacts product density or expansion.

A disadvantage of using this chamber post-extrusion to control product expansion is the high maintenance and operating (utility) costs associated with the hardware. The industry has moved away from this approach to other extruder control measures such as screw speed and vents to control product expansion.

223 What causes my products to be constantly distorted in shape?

Distorted shapes can result from a number of factors.

- If the **moisture level is too high during extrusion**, the final product will usually end up with moisture levels well above those of the melt transition region, resulting in partial collapse of product and shrinkage. This product will also be subject to distortion and clumping caused by a soft tender structure. Correction of this process situation involves decreasing water or steam addition levels.
- **Excessive extrusion temperatures** can also be responsible for product distortion. Elevated temperatures also push extruded products above the melt transition

temperature and result in soft tender products that are subject to distortion when they are cut at the die or during the conveying step into the dryer. Clumping will also be an indication of elevated temperatures. Temperatures can be lowered by decreasing steam addition directly into the product and by lowering extruder barrel temperatures through increased external cooling.

- **Long die land lengths** can indirectly be the cause of product distortion. The longer land length results in increased restriction of the product at the die and this elevates extruder load and increases mechanical energy inputs. The net result is higher product temperatures and distorted products.
- **Aggressive extruder configurations** impart high levels of shear or mechanical energy and result in elevated product temperatures and distortion.
- If possible, **the recipe can be modified** to reduce or eliminate the levels of ingredients that contribute to stickiness. Often this is corrected by avoiding ingredients that lower the melt transition temperatures which bring the final extruded product closer to the rubbery region.
- **Adjusting the number of knife blades** may also have an impact on product distortion. Too many blades result in a slow knife speed and increase the time during which each blade covers an orifice in the die which can then give 'mushroomed' ends on the final product. Too few knife blades will require an excessive knife rotational speed and throw product against the side of the knife hood or take-away system and increase the occurrence of clumping and distorted products.

224 What causes a burnt appearance and taste in extruded products?

The burnt appearance and taste in extruded products is normally the result of processing a recipe at extremely high temperatures and low moistures. Browning can be the result of the Maillard reaction as proteins, in combination with reducing sugars, are processed at high enough temperatures to facilitate it. In some applications, mild browning may be desirable for taste and appearance effects. However, a process that actually results in a burnt appearance and taste (see Fig. 1) is usually not desirable and can be avoided by processing at moderate temperatures and higher moistures.

A burnt appearance can also result from over-drying or drying at too high temperatures. Drying is a process that usually follows extrusion to reduce product moisture to a reasonable level for shelf stability.

225 How can I increase cook in the extruded product without over-expansion?

Many of the techniques suggested in Question 222 earlier also apply to the process question raised here. Other process factors to consider are as follows:

Fig. 1 Sample of scorched product.

- **Addition of more energy** into the process is usually effective in increasing cook levels. However, the product may be likely to over-expand and then it becomes necessary to employ the techniques reviewed in Question 222.
- **Increasing the retention time in the preconditioning step** may also be effective in increasing cook values. Simply extending retention times at elevated process temperatures will improve hydration, swelling and gelatinization of recipe starches. This technique may also result in more expansion requiring that the factors in Question 222 be implemented.
- A simple approach to increasing cook without the accompanied expansion is to **add extensions between the end of the extruder screw and the final die plate**. The added distance that the product must travel after the screw before exiting the die orifice affects the process in several ways. The product is subject to longer retention times which will increase cook. As the product is conveyed (usually in a plug flow) the product temperature may increase slightly due to the extra energy required to 'pump' the product through the longer distance. This increase in product temperature will increase cook and may even require some moderation or tempering by jacketing the die extension in order to circulate cooling water. However, cooling the extension increases product viscosity which could also lead to slower moving and even stagnant material on the inside side walls of the extension. The longer die extension does result in a pressure drop as the product flows towards the final die, which moderates expansion. The length of the extension can be used to control the amount of expansion. Longer extensions result in less expansion.

- **The use of partially or fully pregelled or precooked ingredients** will obviously increase cook levels and reduce the energy input required during the extrusion process. The net result is more cook and less product expansion.
- Depending upon the glass and melt transition temperatures of the recipe components, it may be possible to reduce expansion and still achieve high cook levels by **adding 'excess' water and/or steam** during extrusion. This water will increase product moisture to the point that considerable collapse of the expanded product will occur after the die, resulting in less expansion even at the conditions that will increase cook levels.

226 How can I maintain proper product density?

There are a number of factors that contribute to product density. Understanding and then controlling or manipulating these factors are essential to maintaining target product density:

Recipe contributions
Consistent formulations or proper recipes are key. Those constituents that affect product expansion during processing must be held constant. Starch and functional proteins assist in expansion. Fats and lipids and non-functional proteins inhibit expansion. Grind or particle size also influences product expansion at a given set of processing parameters. All of these variables must be held constant in order to maintain consistent product density.

Thermal energy inputs
Regulating thermal energy inputs such as direct steam injection into the preconditioner or extruder barrel are the most efficient steps for adding or taking away thermal energy. Perhaps the most significant variable is direct injection of extruder steam as this is the highest pressure steam and contains the most energy. Steam addition to the preconditioner is also effective but usually steam pressures are only in the 2–3 bar range and thus contain less energy. Thermal energy can also be inputted by simply increasing process water temperature. Hotter water contains more thermal energy. Another method of controlling thermal energy input, albeit much less effective, is to control barrel temperatures via the circulation of water, steam, chilled fluids or oil through the extruder head jackets.

Mechanical energy inputs
Controlling the level of mechanical energy input is very effective in maintaining product expansion or density. Mechanical energy input is measured by monitoring the load on the main drive motor of the extruder. On many occasions variables such as mechanical and thermal energy input are interrelated with one another and other

Fig. 2 Self-wiping, positive displacement twin screw elements.

variables such as extrusion moisture. Changing mechanical energy input includes adjustment of die restrictions, modification of screw configurations, use of certain tools that adjust restriction to product flow at some point in the barrel. Twin screw extruders are excellent pumps owing to their positive displacement characteristics and self-wiping properties (Fig. 2). Making changes to the screw configurations to increase specific mechanical energy usually involves decreasing transport properties and interruption of conveyance.

Extrusion moisture

Moisture is an excellent plasticizer and has a great influence on product viscosity at various temperatures. If extrusion moistures are not held constant during a process, the product density is not likely to be constant.

Wear or dimensional condition of internal barrel components

If certain barrel components – screw elements or extruder barrel liners – are worn, it is likely that the process will not be stable and thus product density will be difficult to maintain.

227 How can I reduce the cell size in products?

The cell or pore size of an extruded product can be influenced by several factors:

- **Recipe selection**: Large particle size of recipe components will tend to result in larger cell structure. In addition, significant levels of functional proteins and/or fat in the recipe will tend to enlarge cell size.

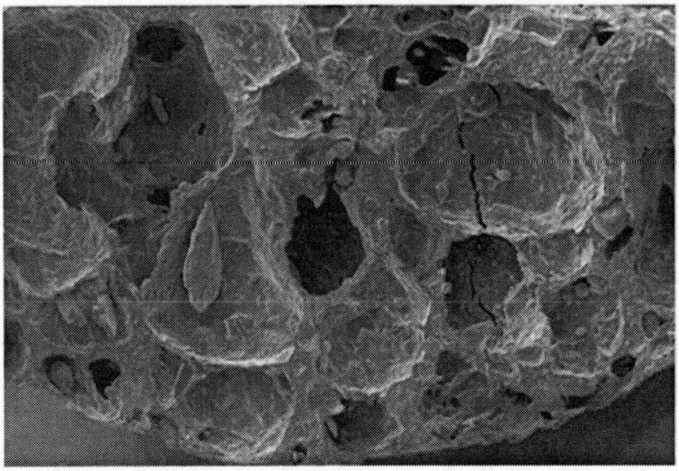

Fig. 3 Scanning electron microscope (SEM) image of cross-section of extrudate with a specific mechanical index of 1 showing a large cell structure.

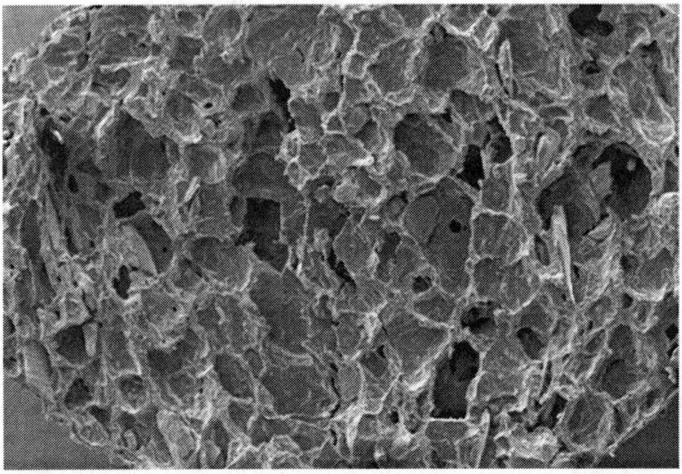

Fig. 4 SEM image of cross-section of extrudate with a specific mechanical index of 1.4 showing a large cell structure.

- **Extrusion moisture**: Adding more moisture during extrusion, either in the form of steam or water, can result in larger cell size.
- **Mechanical energy input**: When the extrusion process is adjusted to increase specific mechanical energy inputs, there is usually a correlating change in cell size (see Figs 3 and 4). As the level of mechanical energy is increased, the size of the cell structure becomes smaller.

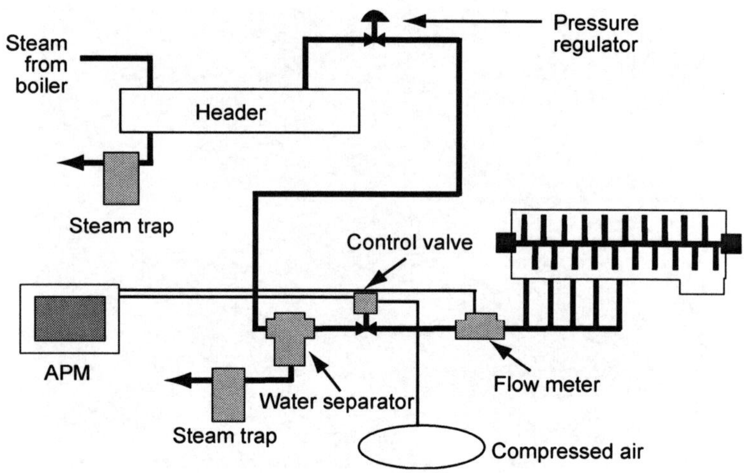

Fig. 5 Example of steam supply plumbing.

228 Why does product moisture and expansion vary when adding steam into the extruder barrel?

Product moisture and expansion should not vary to a large degree during a stable process. Significant variation in these parameters is likely to indicate the presence of condensate in the process steam supply. If the steam supply plumbing does not provide quality saturated steam without condensate, there can be visible variations in moisture levels available to the process. An example of a steam supply harness is found in Fig. 5. Note that condensate is removed in the header device and that further precautions are taken to remove condensate using the water separator.

Another source of moisture variation and product expansion is inconsistent steam flow into the extruder. This may be caused by intermittent plugging of steam injectors due to excessive barrel fill. If the barrel fill is too great, product pressure may momentarily or constantly exceed steam pressure, preventing consistent flow. Steam supply pressure should be verified that it is greater than 7 bar (105 psig) and preferably above 8 bar (120 psig). Excessive barrel fill may also be due to worn components which reduce transport or conveying power. Excessive feed rates and/or die restriction may also contribute to excessive barrel fill.

229 What causes the extruder to leak product along the length of the barrel?

Most extruder barrels are designed in a segmented configuration to allow for replacement of short portions of the barrel and greater flexibility in modification of the setup. Every segment section has a mating surface where it connects to the adjacent section. The mating surfaces provide an opening where product can

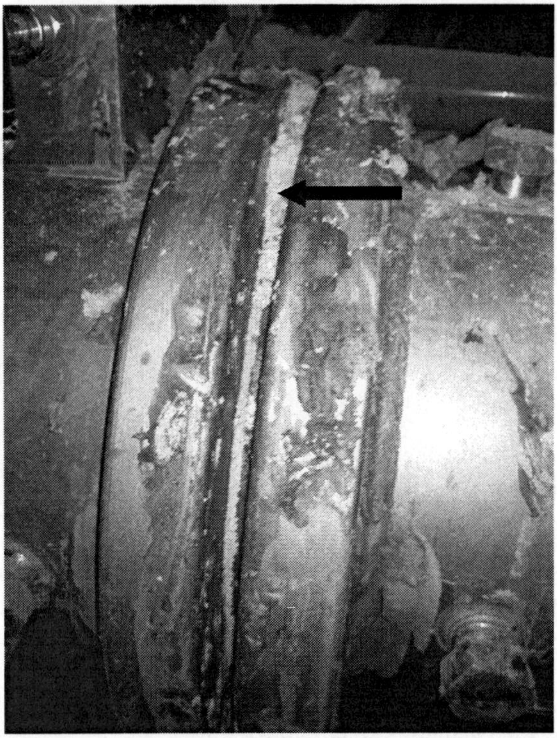

Fig. 6 Arrow indicating product leakage from joints between barrel segments.

potentially leak out if the surface is damaged or marred. If the mating surfaces are not properly cleaned when disassembled there is also opportunity for product leakage. Even if the mating surfaces are clean and properly cared for, product leakage can occur if process pressures are excessive. Occasionally the bolts used to attach the barrel segments to one another can be of inferior quality, resulting in the bolts stretching and creating a gap in the heads, leading to product leakage (Fig. 6).

230 Why do my process steam and water injection levels vary during extrusion?

Constant flow of water and/or steam into the process is critical for stable extrusion and consistent product quality. There are a number of factors that can cause variation in water and steam flow rates:

- **Variations in supply pressure**: The supply pressures to the extruder for water or steam can vary and this will have an impact on the process if steps are not taken in the plumbing assembly to reduce the impact on flow rates. Most plumbing assemblies will include pressure regulators to provide consistent pressures to the flow control devices.

- **Improperly functioning flow control valves**: Water and/or steam flow rates can vary if the flow control valves are not functioning properly. Excessive wear on valve seats or improper tuning can have an impact on flow rates.
- **Partially-plugged lines or injector nozzles**: Debris or particulates in the lines can plug or temporarily plug injection nozzles or orifices resulting in altered flow rates. Cleaning out the lines and installing filter systems will solve this problem.
- **Excessive barrel fill:** If the extruder barrel is too full of product owing to excessive screw wear, too low extruder screw speed or excessive die restriction, the water and/or steam injectors can be partially plugged which will also have an impact on flow rates.

231 How can I read a true product temperature inside the extruder?

Accurate product temperature measurement is required for process control and for food safety reasons when the product temperature is a designated critical control point. Temperature sensors are commonly used to measure product temperature. Where temperature sensors (thermocouples) are inserted through the barrel to contact product flow within the extruder, care must be taken that these sensors are located and mounted in such a way that readings are not hindered or altered by influences other than actual product temperature.

Studies have indicated that the optimum location for a temperature sensor appears to be after the extruder screws and just inboard of the final die plate. Locating temperature sensors along the extruder barrel will not always result in true product temperatures owing to influences from barrel heating or cooling. If the barrel is jacketed, it is likely that the heating or cooling medium that is being circulated through the jacket will influence temperature readings. When temperature readings are taken along the barrel, they can be used for process setpoints when producing a certain product, but are not indicative of the true product temperature.

The following are recommended guidelines for measuring product temperature accurately:

- Placing the sensor in a die spacer or adapter between the end of the screw and the final die is an ideal location.
- The tip of the sensor should be inserted at least 6 mm into the product stream from the side wall of the extruder (see Fig. 7). In some processes, stagnant product builds up along the inside surface and it is important for the tip to extend beyond the material that is not flowing.
- The sensor tip that extends into product flow may need to be protected by a metal sheath to prevent excessive wear.
- It is recommended that where the temperature sensor passes through the side wall of the extruder or die assembly, it is protected from convectional heat and outside influences by an insulated sheath.

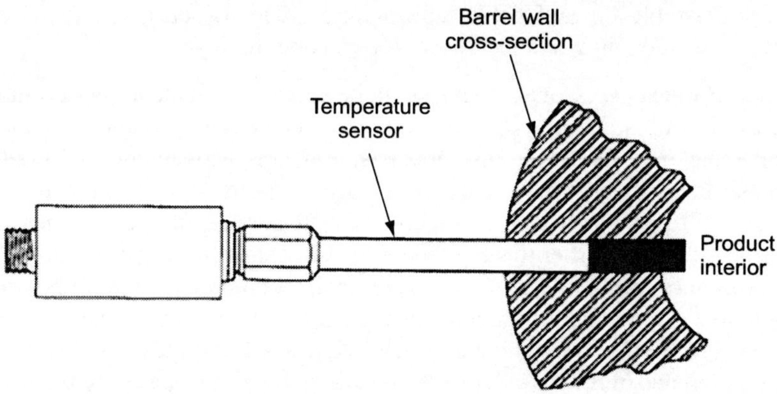

Fig. 7 Correct placement of the temperature sensor.

- More confidence can be put into the resulting reading if two sensors are used. If the readings agree, this gives confidence that the sensors are reading correctly. If the two readings are not close, then it is an indication that one of the sensors may not be functioning properly.
- It is suggested that each sensor be calibrated once a month by using boiling water to check actual temperature accuracy.

232 What is the best method of controlling product temperature?

Controlling product temperature can only be accomplished if you are first able to measure the product temperature correctly. Temperature sensors commonly used to measure product temperature must be located and mounted in a way that readings are not hindered or altered by outside influences. Guidance is given in Question 231.

Now that it is possible to read accurate product temperatures, the actual control of that temperature is possible. Product temperature is controlled by adding to or removing energy from the process. This involves process variables such as steam and water levels added to the product and mechanical energy input factors such as screw speed, extruder configuration, die restriction, and so on. Methods to effect heat transfer in or out of the product through external sources such as cooling water or other heating mediums in the barrel jackets are often used but may be much less effective than direct energy input such as process steam injection.

233 What causes the extruder barrel to move in an eccentric fashion?

There can be several reasons why an extruder barrel can 'move' during operation.

It is not desirable for noticeable movement to occur which can actually cause fatigue in the mounting connections if severe enough.

- **Lack of barrel supports**: Some small detectable movement may actually be expected with some processes. However, as mentioned previously, continual movement may result in fatigue of connection parts. Movement can be reduced or even eliminated by the use of proper barrel supports. Most extruder manufacturers will provide structural support from the frame of the extruder to the extruder barrel in order to supply the necessary strength to the structure.
- **Problematic recipe**: There are certain recipe components that either do not hydrate easily or remain viscous during extrusion and never attain the 'melt' stage. The highly viscous material will often not distribute evenly around the circumference of the extruder screw resulting in voids or partially filled screw volumes. The uneven distribution can result in noticeable barrel movement, especially at high screw speeds. Usually the addition of more moisture, either through the addition of more water or of more steam, can reduce the severity of this problem.
- **Misalignment**: A bent extruder shaft or misaligned extruder barrel can also be responsible for unnecessary movement. Mechanical realignment and inspection of the shaft can provide a remedy.

234 How can a knocking noise be eliminated from the extruder when it is in operation?

The extrusion service engineer will occasionally find a situation where there is a 'knocking noise' coming from the extruder barrel. Although this is not common

Fig. 8 Shaft-centering device for a twin screw extruder.

and not to be expected, it does not necessarily indicate a critical situation requiring emergency action. The severity and loudness of the knocking determines the seriousness of the situation. Some extruders which are slightly out of alignment will emit a knocking noise during the startup phase until product fills the extruder barrel and serves as a bearing to prevent screw/barrel liner contact which is responsible for the noise.

The following actions are required to reduce or eliminate this noise:

- Check the extruder shafts to ensure that they are not bent and straighten them if necessary.
- Align the extruder barrel around the shafts as required.
- Install a shaft-centering device which prevents screw/barrel liner contact and can reduce wear by up to 30% (see Fig. 8).

235 What causes the extruder throughput to decrease over time?

Each extruder has a given volumetric capacity based on the barrel/screw volume and screw speed. If the extruder throughput decreases over time, it is usually the result of worn screw elements and barrel liners. While it is true that worn screw elements and barrel liners will slightly increase the total volume within the barrel, the impact on transport capacity is greatly reduced. As the leading edge of the

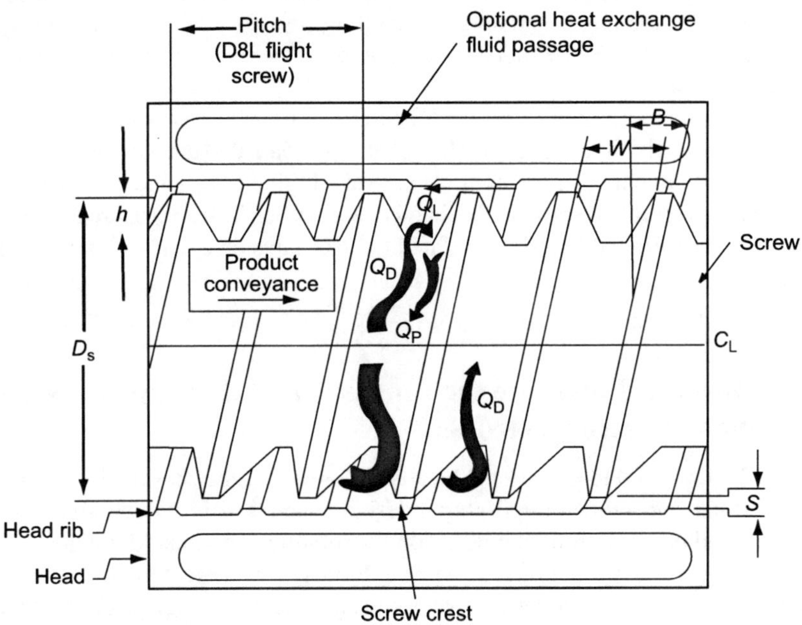

Fig. 9 Product flow patterns between the screw and the barrel wall.

screw flighting is worn away, the tip of the flighting begins to round off and the lead angle of the flighting becomes more flat. Both of these factors greatly increase leakage flow (factor Q_L in Fig. 9) and reduce transport. The increased clearance between screw elements and the barrel wall and the elements themselves creates gaps between flights and the self-wiping and positive displacement characteristics of a twin screw system are diminished. The net result of a decreased transport ability is a fuller barrel (i.e. product fill increases) which requires more power. In addition, barrel fill may increase to the extent that steam injection and even water injection ports may be lost. The loss of energy from direct steam injection increases viscosity and puts a further load on the main drive motor. The combination of greater barrel fill and increased product viscosity creates motor overload. To compensate for the gradual increase in motor load, the dry feed rate is often reduced in order to remain in a safe operating range without overload conditions.

236 Wear on my extruder components is excessive – what can be done about this?

Excessive or above normal wear on extruder components can be due to a number of factors:

- Very abrasive recipe components: High levels of fiber, minerals, or other ingredients that are abrasive will accelerate wear.
- Large particle size of raw materials: The larger the particle size of recipe components, the more 'grinding' or 'melting' the extruder must accomplish to mix and cook the recipe thoroughly in order to move above the glass and melt transition curves.
- Low moisture extrusion: Lower moisture levels during extrusion result in less viscous material in the extruder barrel and this can also accelerate wear.
- High mechanical energy input: Relatively higher mechanical energy input (often expressed as 'specific mechanical energy' or SME, measured in units of kWhr/t) are almost directly translated to wear rates on extruder barrel components.

237 How can I increase the range of ingredients processed on the twin screw extruder?

As briefly discussed in Question 226, a twin screw extruder has the inherent fully intermeshing and self-wiping flighting design characteristics which give positive transport and prevent product buildup on the rotating elements. Many ingredients such as starches and functional proteins exhibit sticky characteristics as they are hydrated and heated. Product characteristics may result in significant buildup in the root diameter of single screw extruders reducing or stopping positive transport. The self-wiping characteristics of a fully intermeshing twin screw extruder prevent excessive buildup. Other ingredients such as fats and oils act as lubricants and hinder the positive transport of these slippery products. The positive transport offered by twin screw elements will overcome this problem and still provide good conveyance.

As the screw elements wear, they lose transport power even with the twin screw design. Keeping the rotating barrel elements in good physical condition and near proper dimensional tolerances will help maintain transport ability and broaden the range of ingredients that can be formulated into a recipe.

238 Who can help me select the proper extruder configuration?

The correct extruder configuration for a particular application is essential. An extruder will be configured for a particular application based on the shear stress necessary for that product. There are three general categories of shear stress: low, medium and high.

Extruders configured for low shear stress applications will usually have deep continuous flighting with good transport capabilities. Extruder barrel lengths are usually quite long. The extruder drive will be set up to turn at relatively low speeds and the barrel is designed not to generate high product temperatures. Products from this category are dense, high moisture and processed at low temperatures. Examples are pasta, confectionaries and some pet food treats.

Applications in the medium shear stress category will typically include aquatic feeds, pet foods and other moderately expanded products. The flight depth is less than that used in low shear stress applications and the screws speeds are greater. The length-to-diameter ratio is less than for low shear stress applications and the extrusion moisture will be lower and product temperatures higher. Examples are pet foods, aquatic feeds and moderately expanded products.

High shear stress applications include products like highly expanded breakfast cereals and snacks. Extruder barrel lengths are usually very short and flight depths are quite shallow. Product moisture is low and product temperatures are high. The supplier of the extrusion equipment should be able to recommend a starting configuration that will be based on experience once they know the application.

239 What is the difference between co-rotating and counter-rotating twin screw extruder designs?

Most twin screw extruders used in the food industry are co-rotating extruders. A co-rotating twin screw extruder has either two left-handed or two right-handed flights, while a counter-rotating extruder has one left-handed and one right-handed flight. It is not possible simply to turn the screws in different directions as desired without physical interference from opposing flightings unless the construction is specifically designed for that rotation. Counter-rotating twin extruders are mainly used as positive displacement pumps for liquid-like ingredients as the channels formed by the flights are closed and this system would never develop enough cook for most food and feed applications. Co-rotating twin extruders have open channels down the flightings and develop much more cook, as is required in the food and feed industries.

Die and knife assemblies in food and feed extrusion: common problems and their solutions

240 What causes extruded products to be misshapen?

A number of factors can contribute to distorted product shape.

- **Recipe constituents**: Many recipes include significant quantities of ingredients that possess low glass and melt phase transition temperatures. If the glass and melt phase transition temperatures are too low, the extrudate will be very soft and tender at the die and thus subject to distortion from the cutting action of the knife blades and from impact on the product shield or conveying device.
- **Extrusion parameters**: High product temperatures and moistures that are well above the melt phase transition temperature curve can also result in very tender product discharging from the die that is thus subject to the same distortion described above. Product melt transition temperatures that are too low for the process conditions employed can also cause distortion, as can moisture levels that are too low.
- **Poor die flow characteristics**: The die design can contribute to product distortion. The effects of die flow can be studied by observing product flow and cutting at the die by using of a strobe light where the light frequency can be varied to 'freeze' or visually slow down the speed of the knife.
- **Knife design and speed**: The cutting surface area of the actual knife blade can have an impact on product distortion. Broad knife surfaces can cover the die opening for longer periods of time, partially obstructing or hindering product flow (Fig. 1). Knife speeds that are too slow also increase the time that the knife blade covers the die opening and this can distort the product.
- **Handling**: The product can be damaged by harsh handling at any stage in the process after extrusion.

Fig. 1 Typical die and knife assembly for directly expanded products.

241 What can cause non-uniform piece sizes?

There are a number of causes of non-uniform piece size:

- Knife blades are not symmetrically spaced.
- Knife speeds are not consistent.
- Knife blades are broken.
- There is too much die open area.
- Die flow design is poor.
- Die openings are partially plugged.

242 How can product expansion be controlled?

Product expansion is generally controlled by a combination of recipe characteristics and process variables.

- **Recipe considerations**: The levels of starch, functional proteins, fiber and fat all influence product expansion. Ingredients that hydrate easily and are largely water soluble will contribute more to expansion. Ingredients that contribute to viscosity and have elastic properties will have a greater impact on product expansion than ingredients that are inert or are not affected by the process.
- **Process energy input**: Increasing thermal or mechanical energy input usually results in more product expansion. Regulation of energy input not only comes from altering the processing conditions, but is greatly influenced by hardware changes such as die restriction and extruder configuration.

- **Extrusion moisture**: The adjustment of moisture levels during extrusion will have a marked influence on product expansion.

243 Why does product appear to collapse or shrink as it leaves the die?

If the product has high moisture and/or temperature levels as it leaves the die orifice and the resulting product is above the melt transition temperature curve for the given recipe, this will result in a product that can 'collapse' at this point. To avoid this shrinkage, the product temperature and/or moisture must be decreased or the recipe must be modified to increase the melt transition temperature.

244 Why are some products thicker on one side than the other?

If thickness or product flow are too great, long-products may become twisted or curved in shape similar to elbow macaroni, but short-cut products will have a wedged shape in which one side of the product is thicker than the other (Fig. 2).

Fig. 2 Example of a wedge-shaped product.

Other process variables that can result in a wedge-shaped product include the following:

- Extrusion temperatures are too high.
- Extrusion moistures are too low.
- Knife speed is too slow.
- Knife blades are too broad.

245 When making a ring shape, what causes one part of the ring to expand greatly and the other not to expand sufficiently?

The problem that is described here usually results from an incorrectly positioned pin in the die orifice (see Fig. 3). Either the pin was not correctly centered during assembly of the die components or the pin was extending past the surface plane of the die (occasionally the die is assembled properly but the pin support is not robust enough and is pushed forward by product pressure during processing).

246 What causes 'tails' on individual product pieces?

'Tails' can be described as small appendages of product jutting out from the main product piece. These small appendages will break off easily during transport and handling and result in excessive fines. The major cause of these 'tails' is improper placement of knife blades in relation to the face of the die. A secondary cause is the

Fig. 3 Distorted ring-shaped product from an off-set pin in the die.

extrusion of some products at improper moisture levels. If the exudate is too dry, it may not flow through the die at the proper viscosity and will tend not to remain in one cohesive piece.

247 Why do some product shapes break easily?

Some product shapes break more easily than others owing to their structural composition or lack of it. Some ingredients like fats and oils reduce structural integrity. Product durability can be improved by modifying the recipe, changing the shape to a more architecturally sound pattern and by giving more cook to the product during processing.

248 Why do product pieces occasionally have a mushroom-like appearance?

The 'mushroomed' appearance usually comes from the extrudate hitting an obstruction or another restriction while it is still in the melt transition temperature stage. The most likely causes are slow-moving knife blades or wide knife blades that cover the die orifice for too long. Similar product distortion could occur if the extrudates strike or impact a hard surface (e.g. a knife or the pneumatic-conveying hood).

249 What causes light and dark streaks on the side of the product?

Light and dark streaks or areas along the surface of a product are usually indicative of non-uniform hydration and/or cooking. Simply increasing the moisture content during extrusion or increasing the screw speed will reduce or eliminate this characteristic.

250 What causes rough surfaces on the product coming from the die?

Rough surfaces on the product are a result of low moisture extrusion or excessive mechanical energy input during processing. Increasing moisture levels during extrusion or reducing the extrusion temperature will result in a smoother product surface.

251 What can be done to prevent corrosion of the die and knife parts?

There are several types of wear pattern that can manifest themselves in the extruder die and knife assemblies. The most common wear patterns are related to normal

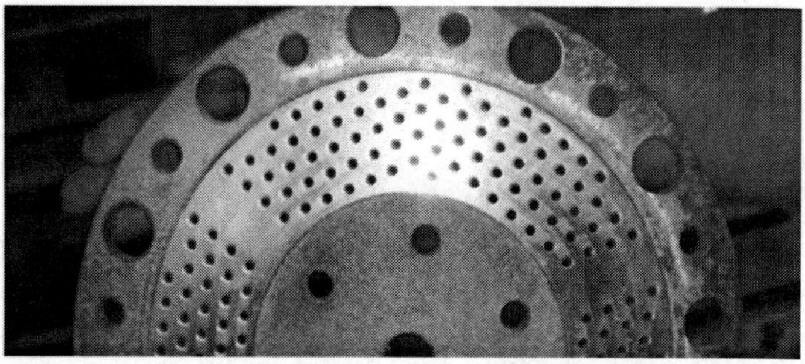

Fig. 4 Elongated openings in a worn die plate.

Fig. 5 Excessively worn knife blades.

abrasion from material flowing through the die orifice (Fig. 4) and from knife blade contact (Fig. 5) with product and with the face of the die. Corrosive wear is different from abrasive wear and usually results from recipes that are low in pH (highly acidic) or have high levels of salt. Selecting the proper metallurgy is the best step to take to reduce or eliminate corrosive wear.

252 Why do some knife blades wear out quickly?

Knife blades can experience excessive wear for several reasons:

- Recipe constituents are very abrasive, resulting in excessive wear.
- Product melt transition temperatures are too high for the process parameters employed.
- The metal composition of the knife blades is too soft for the process conditions.

Fig. 6 Die inserts with gouge marks from knife blades.

253 What causes gouge marks on the surface of the die?

Marks on the die surface will most likely come from the knife blades (Fig. 6). If the blades are adjusted too close to the surface of the die face, they can actually wear grooves into the face of the die. This will often result in broken knife blades as the metal swells and contact between the surfaces becomes more intense. Other causes of gouges, such as tool marks from improper handling of tools are possible but less likely.

254 What is the optimum knife speed?

Optimum knife speed can be defined as the speed at which the correct product length is achievable while not having to operate the knife at excessive speeds. High knife speeds can result in accelerated wear of both the knife blades and the die face. Low knife speeds will probably result in deformed product because the knife blade resides over the die opening for too long.

255 Why does product leak out between the die plates?

Product can leak out between die plates for several reasons:

- If the plates are mounted incorrectly they may not be seated as required resulting in gaps where product, even at moderate extrusion pressures, will leak into the environment.
- Hard residual product may not have been properly cleaned away from between the plates or mounting shoulders.
- Poor handling of die plates during cleaning or mounting can result in burrs along the seating surfaces which will result in improper seating.

256 Why does the extruder continually plug up at the die during startup?

The die openings on an extruder can plug during startup if the extrusion moisture is too low. This phenomenon can be avoided if more water or oil is added during startup mode to reduce product viscosity. It is also possible that some particulates in the raw materials are too large and will plug or partially plug the die orifices. Occasionally an extruder die will plug during startup because hard partially dry residual product remains in the extruder screw from previous production runs. The extruder should be thoroughly flushed with water between production runs to avoid the collection of residual product within the system. This is especially critical if the extruder will be idle for more than 4 hours, as any residue has the opportunity to harden and plug or partially plug the die orifices.

257 What could cause the die plates to be blown off the end of the extruder during a run?

If the die plugs during startup or during a production run to the point that extrusion pressures are excessive, this phenomenon could occur. It may also occur if the extruder operator attempts a startup when the moisture content is too low or if the process water supply is interrupted or lost during production. In some processes where process steam is utilized at high levels, the same can occur if the steam supply is lost or temporarily interrupted. Process applications where the oil or fats are added to the extruder are more subject to excessive pressures when oil or fat flow is interrupted. One mechanical cause of die plates coming off the extruder may be due to using bolts that are too short to attach the die assembly to the end of the extruder. The extruder manufacturer will supply maximum operating pressure data for the system and this should not be exceeded.

258 Why does product flow faster on one side of the die plate compared to the other?

There are a number of reasons why product flow through the die may not be uniform.

The arrangement of the die openings must be symmetrical around the face of the die. Openings that are spaced more closely together in one area of the die face will probably have faster flow or product velocity compared to another section of the die face where die holes are spaced further apart.

Each die orifice must have exactly the same cross-sectional dimensions. Shorter land lengths or different tapers on the infeed side will prevent the product from flowing evenly.

If some die orifices are partially plugged, the product velocity will tend to be slower through these holes as product flow will take the path of least resistance.

259 Why does the product diameter get larger over time?

The diameter of the final product coming out of a given die opening will be larger when it is produced using a worn die than when a newly manufactured die assembly is in place. Usually the simple explanation is that the diameter of the die orifice enlarges as it wears, resulting in a larger diameter product.

260 What is die swell and how do I take it into account?

In simple terms die swell is the increase in product diameter beyond the size or diameter of the orifice through which the etxrudate passed during processing. The term 'die swell' is inaccurate as it is the product that swells or changes in size, not the die itself. The swelling of the product occurs because of the elastic nature of the recipe components. The amount of 'swell' is determined by die design, melt transition temperature of the recipe, and the elasticity and viscosity of the extrudate. Most food extrudates actually swell to a size much larger than the die orifice but 'shrink' back somewhat depending on the recipe properties and processing conditions. Die swell can best be anticipated by mapping the melt and glass transition temperatures for a recipe at various moistures. Extruder temperatures and moistures can then be plotted on the map to determine the degree of expansion or swell and the degree of collapse or shrink at a given set of operating parameters.

261 How can die flow be streamlined?

The die flow can be streamlined in two ways:

- The product flow channel can be streamlined by design. Simply said, the flow channels should have a minimum of corners or sharp turns or radii and use tapers to make the transition from one diameter to another.
- The surface of the product flow channel should be as smooth as possible to reduce the coefficient of friction. Buffing of metal surfaces, lining the flow channels with a material such as Teflon, or coating the flow channel with a material that improves slip or reduces friction are all effective.

262 How can a die be designed to manufacture a product of a particular shape?

Although computer modeling is helpful in designing the die orifice properly in order to achieve a desired shape, experience and trial and error are still invaluable. The science is advancing, but this practice is still an art.

263 Where can I get help with designing new die shapes or with die technology in general?

The best source of assistance with die technology in general is the supplier of your extruder hardware. There are a few vendors that specialize in the die assemblies that are required to make unusually or exotic product shapes and sizes.

264 What is an on-line variable restrictive die?

Some dies are designed to allow the amount of restriction to be varied during processing. Designs to accomplish this practice can vary from a simple bolt that is threaded into a die orifice during processing to an electric servo motor that positions elaborate restrictive devices as required.

Drying extruded food products: common problems and their solutions

265 Why do extruded food products need to be dried?

Water and steam are two components added to a product formulation during the extrusion process to facilitate cooking and forming the final product. This includes products such as ready-to-eat cereals, snack foods, food ingredients, pet food and aquaculture feed. Although required for cooking and extrusion, excessive unbound free water in the finished product will lead to product quality issues and reduced shelf life through microbiological, chemical and enzymatic reactions. Regardless of how well one formulates and extrudes the product, if it is not dried and cooled properly, the end result is tonnes of wasted product and lost profits.

There are many ways to dry products, from laying the product out in arid locations and using the sun to dry the product, through batch process dryers, to highly automated continuous process dryers. A dryer removes water from the product by evaporation. It is designed to handle the evaporated water in the most compact machine using the least amount of energy possible while creating or preserving the desired characteristics of the final product.

The basic factors controlling drying are temperature, airflow, retention time, mixing/spreading, process air conditions, and product physical and chemical characteristics.

266 Why are there so many different types of dryers, and which is the best for drying extruded products?

There are many types of dryers to choose from, including flash, rotary, spray, tower, conveyor, fluid bed, tray and others. In determining the right dryer for a

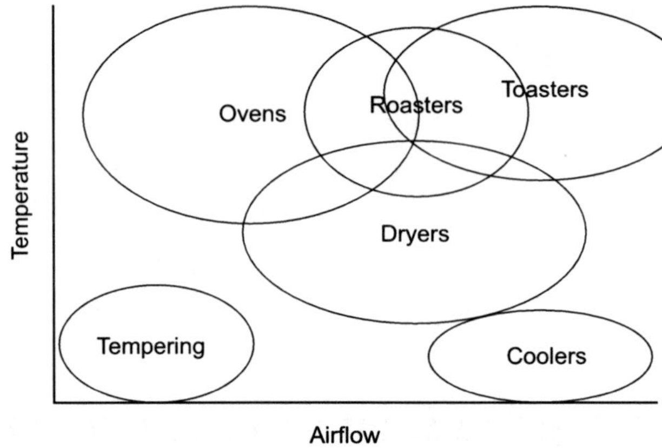

Fig. 1 Thermal processes.

product, we must look at two aspects: (1) what thermal process are we trying to accomplish and (2) how do we convey the product mechanically through the process based on product characteristics? The answers to these two questions lead to the choice of the generic type of dryer. The specific product characteristics, facility requirements and production specifications further lead to a unique dryer configuration within the category of dryer selected.

Many people confuse dryers with ovens in their discussion terminology. However, from a thermal processing standpoint, there are very specific definitions for the types of equipment that use heat and airflow to process the product, which may result in some amount of drying.

The following are the general types of convective thermal processes:

- Ovens, whose primary function is to bring about a chemical or physical change to the product by heat transfer;
- Dryers, whose primary function is to remove moisture through; heat and mass transfer;
- Roasters, whose primary function is to change the flavor profile through heat and mass transfer;
- Toasters, whose primary function is to change the color and possibly the flavor through heat transfer;
- Dry Puffer/Expanders, whose primary function is to change the internal structure of the product through heat and mass transfer;
- Cooler, whose primary function is to remove heat from the product.

By making adjustments to the operating parameters of air flow, temperature and air exchange (exhaust), each type of equipment can sometimes do the job of another, but usually there is a 'best' type of equipment for the required process. The types of thermal process equipment are depicted in Fig. 1.

Although some extruded products are processed on rotary, fluid bed or batch dryers, the vast majority are run on conveyor dryers (see Fig. 2). There are

Fig. 2 Conveyor dryer.

numerous conveyor dryer configurations from which to choose. The simplest configuration in terms of the conveyor beds is a single pass dryer, which has only one conveyor bed. Multiple pass dryers have two or more beds stacked one over the other. The product drops from one bed to the next as it progresses through the dryer. Similar to the multiple pass dryer, the multiple 'stage' dryer is configured with one conveyor after another rather than one above the other.

The choice between the various conveyor bed configurations as well as of the length and width of the beds depends on a number of factors. A fragile product that does not stick together can be handled in a single pass dryer. A product that tends to clump together when wet should be dried on a multiple pass or multiple stage dryer. This allows the product to run along at a shallow bed depth first and to be stacked gradually deeper as the product dries. It is also less likely to clump together.

Equally diverse and equally important is the choice of airflow configuration. Air can go up or down through the beds, it can go through one bed before being reheated or it can go into multiple beds in series before being reheated. The choice depends on the product to be dried.

A conveyor dryer is typically zoned to allow different temperatures, airflow rates or humidity levels at different points in the drying curve. The number and location of zones are additional variables in the dryer configuration.

267 What is the best way to improve moisture uniformity in the final product?

A large number of issues can affect moisture uniformity in the final dried product.

The first is to verify that the moisture content of product entering the dryer is consistent. It is also critical that each piece of product travels along the same process path as it turns from a soft wet extrudate into the stable dry end product. If there are any differences in conditions along this path, inconsistent product moisture may result. The dryer is a point at which process conditions commonly vary and cause differences in moisture levels. Variation in the drying is caused by, but not limited to, inconsistent bed loading, uneven airflow, uneven temperature distribution, product clumping and uneven product retention time.

The simplest way to evaluate drying uniformity is to imagine what a single particle 'sees' as it goes through the dryer. On average, do all particles 'see' the same quality and quantity of process air for the same amount of time? If so, the final product will be uniform. If air flows only up through the bed, the bottom of the bed will be over-dried compared to the top of the bed. If hot air and cold air are not mixed properly, there will be hot and cold spots in the dryer and hence dry and wet particles. If airflow is not uniform across the beds, there will be a moisture gradient across the bed. Airflow uniformity can be affected by dryer design, product characteristics or bed loading.

Air will take the path of least resistance, which is an area on the dryer bed that is shallow compared to the rest of the bed. The greater volume of air passing through this product results in dryer and hotter product. There is a direct correlation between product temperature and moisture content.

Inconsistent bed loading is one of the leading causes of moisture deviation and is one of the easiest problems to detect and correct. A bed of product creates a pressure drop or resistance to airflow, through the product. As the bed depth increases, the resistance increases. Areas where the bed depth is shallow will see an increase in airflow volume and velocity. Conversely, the deeper sections of the bed will see reduced airflow. Greater airflow results in greater heat and mass transfer, causing the product to dry faster than the product in the deeper sections of the product bed. Relating this back to the drying curve, the product in the shallow bed experiences a compressed drying curve compared to the other product. Since there is a strong correlation between product temperature uniformity and moisture

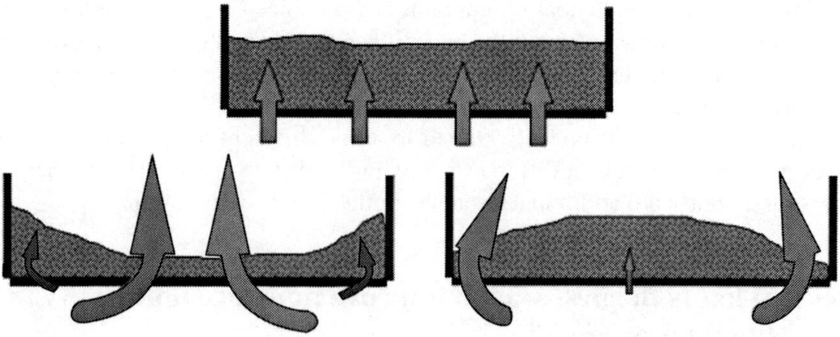

Fig. 3 Correlation between product depth on the bed and airflow.

uniformity, product moisture uniformity can quickly be checked by measuring the discharge temperatures with a handheld infrared (IR) temperature sensor. The arrows in Fig. 3 depict airflow as it relates to product depth on the bed; larger arrows represent more airflow and a greater degree of drying.

268 Why is controlling process air humidity important?

When relating the humidity of the recirculation air of a drying system to dryer performance, the important idea to keep in mind is that the system's humidity and the dry bulb temperature are two critical external factors which govern the speed with which a product will release its moisture to the outside environment. In general, the 'drier' the air, the greater the driving force for drying. Therefore, one would think that a perfect drying scenario would utilize very high temperature air and very low humidity. In practice, the problem with this theory is that most products cannot withstand elevated temperatures. In addition, the extra energy needed to increase and maintain dryer temperatures, coupled with the higher initial cost of larger exhaust fans, ductwork, and associated systems decreases the overall efficiency of the dryer and results in appreciably higher energy costs.

Additionally, some products dried at these accelerated rates will often seal moisture in the interior of the product because the exterior was dried too quickly, subsequently forming an exterior seal. This is sometimes referred to as case hardening. If the surface is dried faster than moisture can diffuse from the interior of the product, surface checking can occur. Checking is small cracks in the surface of the product, which makes the product more brittle. This is a major concern in the production of snack pellets. Considering the above, the overall goal in setting up a dryer is to optimize the drying rates without sacrificing product quality and energy usage.

The humidity level within the dryer zones can be controlled on a continuous basis by installing a humidity sensor in the air stream of the drying zone. A feedback controller can then be used to adjust the damper positions automatically for the make-up and exhaust levels of the zone. The humidity controls will adjust the exhaust of the dryer based on evaporation loads in the dryer. During periods when the dryer is empty, the exhaust can be limited to minimum values. The energy that is saved by limiting exhaust levels will result in a lower energy cost drying system.

During operation, implementing humidity controls will keep the dryer operating at optimal humidity levels. For a dryer designed for a feed rate of 20 000 lb hr^{-1} (9070 kg hr^{-1}) and an evaporation load of 3700 lb hr^{-1} (1680 kg hr^{-1}), an increase of 2% relative humidity in the recirculating air flow of the dryer will decrease yearly heating energy costs by approximately US$50 000. This is based on fuel or heating costs of US$0.8/therm (US$0.0076 per MJ). This will also decrease the exhaust airflow by over 35%, which will reduce any expenses associated with treating the exhaust air.

Table 1 Impact of moisture content in product coming into the dryer

Product into dryer (kg hr^{-1})	Product moisture into dryer (%)	Product out of dryer (kg hr^{-1})	Product moisture out of dryer (%)	Water evaporated by dryer (kg hr^{-1})	% Change in water evaporated by dryer
1200	25	1000	10	200	–
1286	30	1000	10	286	43
1266	25	1000	5	266	33
1357	30	1000	5	357	79

269 What keeps a dryer from reaching and holding the operating temperature set point?

One variable in the moisture removal equation is heat. In the early zones of a dryer, water is being removed very rapidly and there is a great deal of evaporative cooling that taxes the capacity of the heating system. Since this high rate of evaporation is keeping the product relatively cool, we can use high temperatures to accelerate the drying process with no real possibility of damaging the product. Therefore, it is critical that the heat sources have the capability to hold and possibly exceed the zone's set point temperature.

Most processors are thought of in terms of total product produced per hour from a dryer. Dryer manufacturers look at the process in terms of total water evaporated per hour by a dryer. The mass of water evaporated drives heat source sizing, airflow and exhaust. How much impact can changing moisture content by a few percent have on dryer performance? Table 1 demonstrates the increased 'work load' on a dryer caused by simply changing incoming or outlet moisture contents.

Most dryers have a life expectancy of over 20 years. Over the course of a dryer's life, it will be likely to run a variety of products, many of which are introduced after the dryer is installed and under different operating conditions.

It is interesting to note that the output is the same in each of the above examples, but the workload increases dramatically. If a dryer is not drying product at the same capacity as in the past, a good starting point is to look at the designed 'evaporative capacity', and see if the moisture content in the product has changed. The dryer heat source would have been designed for a given evaporative load, with a built-in safety factor. If the evaporative load exceeds that of the heat source, the burners will be firing 100%, or the steam control valve will be 100% open, and the total energy will not be sufficient for the product's requirements. Assuming the product formulation is correct and cannot be changed, the only solution is to add additional energy to the dryer by upgrading the burners or adding steam coils.

270 Why can mold grow on an extruded product in a sealed package or container when the moisture content is at the target 'safe' level?

Condensation occurs by two mechanisms. First, the heat that is stored in the product continues to drive moisture to the surface of the product, where it is released into the confined surrounding air. Second, as the air cools, it becomes saturated with water vapor, reaches the dew point and condenses its moisture onto surrounding surfaces.

As an example, at 190 °F (88 °C) one pound (454 g) of air can hold over one pound (454 g) of water as water vapor in the form of humidity. In volumetric terms, this is over 16 ounces (474 ml) of water. When reduced to 70 °F (21 °C), this same pound of air can only hold one-quarter of an ounce (7.4 ml) of water. Where does the remaining water go? The answer is on to the product, if not cooled properly. Even though the overall average product moisture content and water activity level may be too low to support mold growth, the localized water activity level on the surface of the freshly wetted product will be high enough for mold or pathogen growth.

As a general rule, product that is cooled to within 10–15 °F (5.5–8.3 °C) of the ambient storage air temperature will not have a condensation problem. In some areas where ambient air is particularly hot and humid, process cooling air must be mechanically cooled to ensure the final product is adequately cooled.

271 What causes product clumping and how does it affect product quality?

Most products produced in an extruder are relatively starchy. As starches or sugars dry, they tend to form a glue-like bond on the surfaces between pieces. When these bonds form, surface area from which moisture can be evaporated is reduced.

Dryers rely on maximum product surface area being exposed to the process air to promote efficient and consistent drying. When products cluster together, the available surface area is limited and moisture is more difficult to remove. This results in inconsistent product quality and slight decreases in capacity.

The product on the outside of the clump is exposed to more airflow enabling it to undergo more drying while the product on the inside of the clump remains wetter. If the clump continues to exit the dryer discharge, the individual pieces will have been subject to different drying conditions through the retention of the dryer. This is a large source of moisture inconsistency from piece to piece.

To ensure uniform drying, these starchy surfaces must be broken during the drying process. In a multi-pass or multi-stage dryer, as the product clump falls on to subsequent beds during transfer, it is broken apart, allowing individual pieces to be exposed to adequate drying air. For products that do not tend to break apart during the fall, a mechanical lump breaker may need to be added.

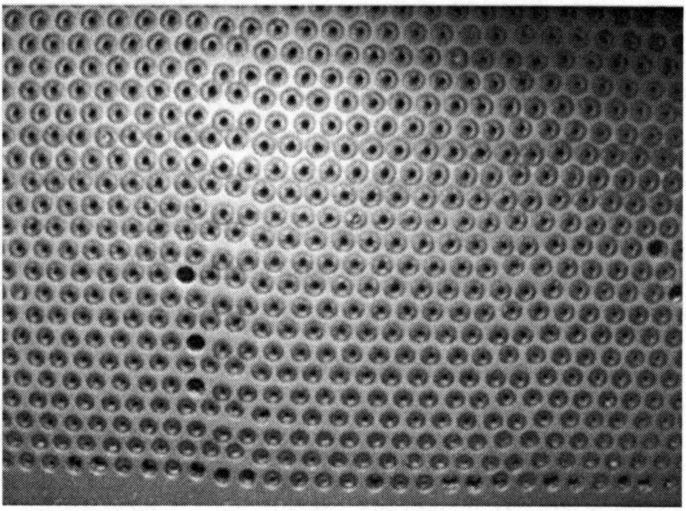

Fig. 4 Bedplate with approximately 50% of the open area clogged.

272 How does sanitation affect dryer performance and product quality?

All drying/cooling equipment built today regardless of the industry in which it operates should be designed with sanitation as a prerequisite. Sanitary design includes eliminating ledges within the dryer where product can become trapped and degrade, peaking the floors of the dryer to facilitate easy clean-out, designing rounded edges instead of flat surfaces to ensure product cannot become trapped and providing access to promote more frequent and consistent cleaning and general upkeep. An unsanitary dryer could easily lead to cross contamination between products, product quality issues and pose a fire hazard.

Since air is the sole vehicle for moisture removal within a dryer, restricted airflow relates to restricted drying. Clogged perforations do not allow airflow to proceed uniformly through the product bed, leading to moisture variation within the discharged product. In severe cases, clogged perforations can significantly increase the pressure drop through the product bed, lessening the volume of air that is able to be circulated by the fans. Figure 4 shows a bedplate with approximately 50% of the open area clogged.

To address these issues, a proactive sanitation protocol needs to be established and followed as part of the plant operation. The bedplates should be inspected and cleaned regularly to re-open the clogged perforations. In extreme cases, productivity increases of as much as 25% have been achieved after a good clean. Additionally, if bedplates consisting of round holes are replaced by bedplates with slotted perforations, maintaining clean perforations is more easily achieved. This is due to the self-cleaning action of the product as it slides across the surface of the bedplates as the product is dumped.

273 What could be the potential cause of a sudden dramatic decrease in production capacity?

For a dryer that consistently runs at a steady production rate, something substantial must have changed to cause a dramatic reduction in production capacity. For a given retention time, a reduction in capacity will be related to a reduction in airflow, temperature or exhaust venting. The dryer must be inspected to determine what has changed to cause this reduction and this cause is likely to be mechanical in nature. In order to control the process parameters in a conveyor dryer operation effectively, all aspects of the mechanical operation of the dryer must be functioning correctly. Air must be forced through the product bed and not allowed to leak by the sides. The operators should inspect the dryer internally to ensure that all internal panels and doors that are used to seal or direct air are in the proper position. If these panels or doors are not correct, air may be allowed to by-pass the product and result in inefficient operation of the dryer. Recirculation and exhaust fans need to be checked to ensure they are actually operating and rotating in the correct direction to remove the amount of air required. Exhaust ducting needs to be inspected to verify that nothing is obstructing the airflow.

274 What causes the energy use of a dryer to increase for a given product over time?

The cost of energy in the dryer is the most obvious cost of drying. You need only look at the large burners in the dryer or at the boiler providing steam to the dryer to realize that money is being spent to remove moisture in the drying operation. This energy is essentially needed to cover four heat loads in the drying operation. They are:

- Warming up the product.
- Evaporating water from the product.
- Heating the make-up air.
- Losses.

Since extruded products leave the extruder at a high temperature, very little energy is used in warming up the product. Most of the energy in the drying operation is used to evaporate water. The energy required to change liquid water into water vapor (latent heat of vaporization) is about 2350 kJ kg^{-1} (1000 BTUs $pound^{-1}$) of water evaporated (the value varies very slightly with temperature). This value cannot be changed n typical dryers, as this is a thermodynamic fact for evaporation at ambient conditions.

A processor may be able to reduce the energy consumed in evaporating water by reducing the amount of water that the dryer must remove. As discussed previously, inlet and outlet product moisture can have a significant effect on the drying load in a dryer. Even a small increase in inlet moisture from 24–27% on a wet weight basis increases the load on the dryer by 27%. Naturally, the inlet

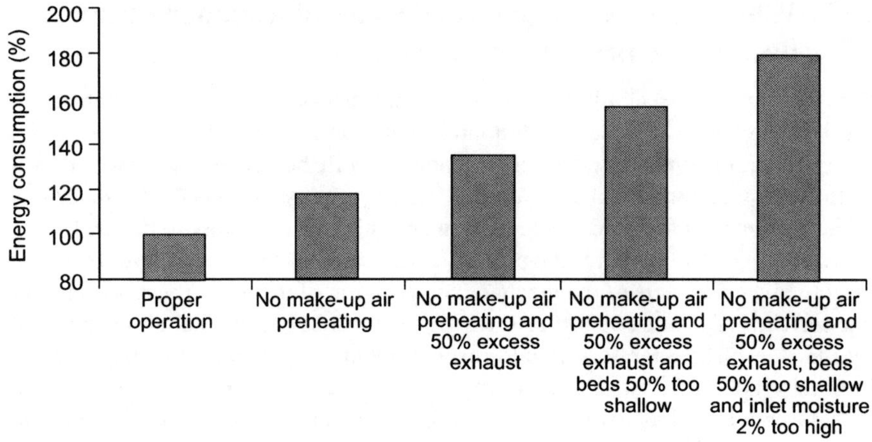

Fig. 5 Calculated energy consumption for an extruded product dryer.

moisture must be driven by product quality considerations, but typically all extruder operators at a given facility do not operate the extruder the same way. If one shift runs the extruder 'wetter' than the next shift, they are wasting energy by adding excessive water to the product only to spend money in the dryer to remove the water.

The third heat load category, heating the make-up air, is typically the second highest heat load. However, in a poorly operated or poorly designed dryer, this heat load can be the largest. This is the area where the dryer operator can have by far the most impact. The dryer needs a steady supply of dry make-up air. Without this influx of air, the process air in the dryer would become saturated with moisture and unable to dry the product. The expertise is in knowing how much air to bring into the dryer. Excessive make-up air will lead to excessive energy consumption. This is because the make-up air enters the dryer at a relatively low temperature and after picking up water in the dryer, eventually leaves the dryer at a higher exhaust temperature. Clearly, energy is consumed in heating the make-up air to the exhaust temperature. It is important to understand that the exhaust air in the dryer is basically the make-up air leaving the dryer along with the evaporated water. To reduce the amount of make-up air in most dryers, the exhaust must be reduced, either by closing the exhaust damper or turning down the exhaust fan.

The temperature of the exhaust and make-up air also clearly affects the energy used to heat the make-up air. If the make-up air is preheated before being introduced into the dryer, less energy will be used by the dryer to heat this air. In an extruded product dryer, preheated make-up air can be drawn from the cooling operation. Spent cooling air used to cool the product is still relatively dry and can be used as make-up air for the dryer. If spent cooling air is not available, preheated make-up air can be used from the residual heat of other processes, such as boilers, or from heat exchangers and heat recovery systems.

Improper dryer operation, such as excessively shallow bed depths (short retention time in the dryer) or internal air bypasses can lead to high exhaust temperatures, thereby increasing the energy used to heat the make-up air to the exhaust temperature. Figure 5 illustrates the significant impact that heating the make-up air and dryer operation can have on the dryer's overall energy consumption.

Unless the dryer is very poorly insulated, losses from the dryer should be low. However, if the dryer design has large areas of uninsulated or poorly insulated surfaces, thermal losses from the dryer could also be costly.

275 How do I predict the drying time required?

Predicting the required drying time of a product, especially a new product, can be a formidable challenge. There are several ways this goal can be accomplished:

- Experience: Based on previous process requirements for drying a similar product, a set of operating parameters can be established which may need to be adjusted slightly based on the current product's particular properties.
- Models: Models for predicting drying times have been developed based on product characteristics such as moisture level in the dryer, product density, product size and shape, and the required level of moisture removal from the dryer. Most reputable dryer suppliers will have developed these programs which are then further improved and refined over time.
- Computer control systems: Mathematical models can be loaded into the control system for the extruder and dryer. These allow the process conditions of the extruder to be communicated to the dryer, which in turn predicts and accomplishes the correct drying time. Obviously temperature and air flow are also critical drying parameters that can be varied as needed to reach the final product moisture.

276 How can I measure the moisture content of the product?

There are a number of methods for measuring the moisture content of a product. The simplest and most time-honored method involves weighing a product as it is heated under controlled conditions and determining at what point the weight of the product is no longer changing due to total evaporation of moisture. The difference in weight of the wet product and the dry product divided by the wet product weight is the moisture content expressed as a percentage. Various instruments have been commercialized to measure moisture content. These include those that operate on conductance principles and those that use microwaves and near-infrared technology. The standard against which most moisture measuring methods are calibrated involves weighing a wet sample into a small aluminum dish which is then placed in a forced air oven at a specified temperature and time. At the end of the specified time, a lid is placed on the small dish and the sample is allowed to cool briefly in a desiccant container before reweighing.

277 What impact do drying and associated processes such as toasting have on the sensory quality of starch-based extruded products?

Drying processes, especially high-temperature drying or toasting processes, have rather complex impacts on the sensory qualities of products which are not well understood. Some products will actually expand further in the drying/toasting steps owing to their elastic properties which enable them to trap escaping moisture (in the form of steam). Other products may shrink during the same process owing to their recipe constituents and moisture content. This expansion or shrinkage can affect textural attributes and mouthfeel. In addition, color and odor properties can be affected. Many snack products and ready-to-eat cereal products rely on the drying step to develop the recognized or expected mouthfeel and crispiness which is only really achievable by reducing the moisture content to very low levels.

278 If the moisture content of the final product exiting the dryer is not correct, should I make changes to the process conditions in the extruder or the dryer so that acceptable moisture contents are reached?

When product moistures are not within specified ranges, it is necessary to make the proper adjustments to the process in order to achieve the desired goals. Too low moisture levels can have a negative impact on the product sensory attributes and other technical qualities. If the moisture levels are too high, the shelf life of the product can also be compromised. Because the moisture level during the extrusion process is critical to many of the technical qualities of the product, it is important to maintain the proper moisture at that point along the process flow. Adjustments to compensate for final product moisture should be made at the drying step rather than the extrusion step. Lowering extrusion moisture in order to reach desired moisture in the final product moisture and to attempt to save on drying costs is the wrong solution. Studies quickly point out that decreasing the moisture during extrusion will often actually increase the overall processing costs even though drying costs are reduced. In addition to alterating the product's technical qualities,

Table 2 Comparative effect of lower extrusion moisture on costs

Extrusion moisture	23%	21%
Mechanical energy costs (US$/tonne)	1.51	2.02
Thermal energy costs (US$/tonne)	1.66	1.57
Maintenance costs (US$/tonne)	1.29	1.59
Amortization (US$/tonne)	0.50	0.50
Interest (US$/tonne)	0.36	0.36
Drying costs (US$/tonne)	4.87	4.29
Total operating costs (U$/tonne)	10.70	10.82

lower extrusion moisture will translate into higher energy costs and maintenance costs for the extrusion process (see Table 2).

279 I have seen both vertical and horizontal dryer designs. Which is best?

Which dryer design is the best depends upon your food process requirements and the space available. If either dryer design can process your product equally well, then the available space becomes a factor. Vertical dryers have a smaller footprint but require more height. Conversely, horizontal dryers have much lower height requirements but occupy a larger footprint. In general, vertical dryers handle the product more harshly owing to the heights involved and the cascading flow through the dryer. If your product has a fragile structure or an intricate shape, the horizontal dryer is the design of choice.

Practical considerations, rules of thumb and solutions to the most common problems in food and feed extrusion

280 What should an operator know about assembling an extruder?

Before startup can occur it is necessary to assemble the extruder. The procedure for assembling an extruder will depend upon the manufacturing style, model and its specifications. As discussed earlier, there are several different types of extruder available. Each extruder will require a special kind of assembly setup. Assembling a single screw is less complicated compared to assembling a twin screw and is relatively easy, whereas with a twin screw a lot of attention needs to be paid to the screw configurations. Similarly, an extruder which has a one-piece screw and barrel is much easier to assemble than an extruder which has several parts. As a rule of thumb, all parts of an extruder should be laid out systematically for quick and orderly assembly. All parts should be checked visually during assembly for defective or excessive wear.

281 Are there any general points that should be taken into account in extruder assembly?

- Check the bolts that secure the inlet head to the yoke. Make sure all bolts are secure and properly in place.
- Check the bolts that fasten the extruder barrel clam-shell together to make sure all bolts are in place and properly secured.
- Check the head supports to make sure head support pins are properly in place and head supports are properly adjusted.

- Check the bolts that fasten the die to the discharge of the extruder to make sure they are installed and properly secured.

282 What utility checks need to be performed before turning on an extruder?

There are pre-startup checks which should be performed before turning on an extruder, which include both equipment and utility checks. Utility checks are listed below:

- **Electrical supply**: Check the main power supply switches and the main DC drive switch to make sure they are on.
- **Steam supply**:
  - Check the preconditioner steam injection to make sure that supply valves are open and the required steam pressure is on. Open the injection valve to flush condensate from supply lines.
  - Check the barrel steam injection to make sure that supply valves are open and the required steam pressure is on. Open the injection valves to flush condensate from the supply lines.
- **Water supply**:
  - Check the water supply to the pumps that supply water to the preconditioner and extruder barrel injection to make sure the supply valve is open.
  - Check the cooling water supply to the hot oil heat exchanger (only if the barrel is heated by oil) to make sure the supply valve is open.
  - Check the cooling water supply to the extruder barrel sections to make sure the supply valve is open.
- **Air supply**: Check the main air supply valve if air motors are being used.
- **Thermal fluid system**: Check the thermal fluid system to make sure the supply and return lines are connected to the proper zones.
- **Instrumentation and electronics**: Check the thermocouples to make sure they are installed and leads are connected to the proper zones.

283 What is the typical startup sequence for an extruder?

A start up sequence for a typical extruder is outlined below:

- Check the diverter spout. Make sure it is set to bypass product from the extruder barrel.
- Start the mixing cylinder and set to approximately medium speed (160–180 rpm).
- Start the feeder. Set it to a low rate (8–10 rpm).
- Wait until product begins to flow down the down spout, then adjust the desired steam flow to the cylinder.
- Start the water pump, which delivers water to the mixing cylinder and adjust to the set point.

- Check the consistency of product coming out of the cylinder when the ingredient is warm and moist or at the desired consistency.
- Start the water pump which delivers water to the extruder barrel and adjust to a set point.
- Start the extruder and turn the screw at a slow speed.
- Divert the material into the extruder barrel.
- Over a 3–5 s time span, turn the extruder up to the desired speed.
- Watch very closely for motor overload and for excessive pressure in the cone section. If either of these are observed, one of the following has most likely occurred:
  - Too much feed is being fed into the extruder barrel. Correct it by bypassing feed intermittently and reduce the feeder rpm.
  - There is not enough moisture in the extrudate. Adjust it by increasing water flow to the extruder barrel.
  - There is a blockage in the die. Correct it by diverting the material flow into the extruder immediately and begin shutdown procedures.
  - If overloading does not occur and the product is flowing out of the die, check the extrudate for moisture. If very wet, begin increasing the extruder rpm or reduce the amount of moisture to the extruder.
- If steam injection is required, adjust the steam injection rate to approximately 0.1 or 0.2 kg min^{-1}.
- When product appears to be of proper consistency to cut, start the knife drive.
- Adjust the feeder settings, moisture and temperature settings to achieve your desired product.

284 What is the purpose of the startup procedure?

The purpose of the startup procedure is to bring the extruder into operation and to a stable condition as quickly as possible. In some cases raw material is very expensive and by delaying the startup, raw material can be lost. In most cases under- or over-processed material can be mixed back into the raw material. But in some cases it is not possible and this will increase the cost of the operation. Before we can start feeding the ingredients to the extruder, the system should be brought as closely as possible to stable operating conditions.

285 What are the critical points which can help to insure quick and steady startup of an extruder?

- Fifteen minutes prior to startup, set the zone control temperatures to the desired setting and start the heating units.
- Extruder barrel steam injection may be used to assist in preheating the extruder barrel to the proper temperature.
- Two minutes prior to startup, start the lubrication pump.

286 Where should I add my ingredients during extrusion processing?

All the ingredients have an appropriate place for addition into the extrusion system. Not introducing ingredients at the proper place may have an adverse effect on the final product and may result in poor and inconsistent quality. The following are recommended places to add solid and liquid ingredients.

- **Dry ingredients** are usually metered into the preconditioner from a live bin or by a feeding screw. If no preconditioner is being used, these dry ingredients can be directly metered in to the extruder barrel.
- **Liquid fat** can be mixed with the dry ingredients in a mixer. Other options are metering directly to the preconditioner or into the extruder barrel (only twin screw) depending upon the product and the desired specification. Liquid fat can also be blended with fresh meat in premixing if the meat is part of the recipe.
- **Emulsified fresh meats** can be metered into the preconditioner or directly into the barrel of a twin screw extruder as part of the dry feed rate.
- **Process water** should be metered into the preconditioner or extruder barrel using manual or automated flow meters and valves.
- **Process steam** should be injected directly into the preconditioner or extruder barrel using a flow meter and valves.
- **Oil or other liquid coatings** can be metered into the coating machine and controlled by a product weighing device. Oil can also be added to product using a batch enrobing mixer.
- **Powder** used for enrobing should be metered into a continuous coating machine from a powder feeder and controlled by a product weighing device. Powder can also be added to product using a batch enrobing mixer.

287 Why do I need a preconditioner?

During gelatinization, the starch granule absorbs water, swells and loses its crystallinity; while in dextrinization, which is favored by extrusion at lower moisture contents, the starch granule is torn apart physically. Both processes cause the starch to become more readily digested. However, dextrinized corn flour has greater cold water solubility. Where maximization of gelatinization is desired, time is required for the added moisture to equilibrate with the cereal fraction before extrusion. Wetting finely ground cereals can be difficult if the dry flour is added directly at the extruder feed inlet with an accompanying stream of water. It has become common to premix and preheat feed stocks in a preconditioner prior to introduction into the extruder. Preconditioners are less expensive to build and have retention times up to three times that of the extruder. However, care must be taken to avoid build up of gluten on the preconditioner paddles when processing wheat flour. Pressurized preconditioners can precook the cereal feedstock before introduction to the extruder. Most manufacturers of cooking extruders now offer steam preconditioners. Similar precooking units are sometimes installed ahead of pellet mills.

Heated non-condensable gases (entrained air) and excess steam can cause serious surging problems when cooking and forming unit operations are both conducted in the same extruder barrel. Precooking in the conditioner reduces these problems because hot feedstock is introduced into the extruder at atmospheric pressure. Some manufacturers have achieved the same effect by installing a vent to depressurize cooked feedstock in the extruder barrel as it enters the final forming section. In some operations, a vacuum is pulled on the vent to increase evaporation of moisture and remove volatile compounds. Also, additional heat sensitive ingredients can be added downstream into the cooked feedstock.

288 What causes an extruder to wear?

A typical extruder can run anywhere between 200 to 20 000 hours before parts are worn enough to be replaced. The life of the extruder screw depends upon the raw material being extruded and several other factors, like moisture, conditioning, configuration, fat content, metallurgy and misalignment.

- **Raw material**: Different ingredients have a different effect on wear of an extruder. Sugar and salt will have more effect on wear than other ingredients. Also coarsely ground ingredients will have more effect than finely ground ones.
- **Moisture**: Higher moisture contents decrease the extruder wear. Direct expanded products with a moisture of 12–16% can wear the parts in 350–450 hours. Products with moisture of more than 25% can increase the life of an extruder screw by more than 10 000 hours.
- **Preconditioning**: Preconditioning can more than double the life of an extruder screw.
- **Configuration**: Intensive use of shearlocks and reverse flight in a configuration will reduce the free volume and will have an adverse effect on extruder wear.
- **Oil content**: Oil content in a recipe works as a lubricant and reduces the wear of the screw. Therefore a high oil content will increase the screw life.
- **Metallurgy**: Extruder parts made from extremely hard alloys will have more life than parts made from a cheap source.
- **Misalignment**: Metal-to-metal is the major cause of extruder wear. This will cause scour marks and sharp edges on screw flights. If you still have knocking noises after the adjustment of a misaligned machine, look for non-parallel screw flights or material trapped between screws. If you still have the problem then suspect bent shaft or bearing housing problems.

289 What is the best method for cleaning the extruder and screws after the run?

Provisions differ for cleaning the extruder screw. In most segmented-barrel single-screw extruders, the barrel sections unbolt and pull off piece by piece. However, occasionally small models are built with a clam-shell design that enables splitting and swinging, opening the two halves of the barrel. The smaller twin-screw extruders are built either with a clam-shell barrel consisting of segments that are bolted together or with solid barrel sections that are pulled off one at a time. One twin-screw extruder manufactured in France includes a railing for pulling the entire barrel assembly off the screws. Pulling the individual worm segments from the shaft may require considerable effort on either single- or twin-screw machines because of product burn-on. It is generally best to organize operations that allow sustained extrusion runs, especially when producing animal foods. However, ease of disassembly/reassembly and cleaning may be a significant factor in selecting extruders for specific operations.

290 What are some rules of thumb that can be used when developing a formulation for extrusion?

It must be remembered when developing a formulation for extrusion that some kind of cohesive network is needed to keep the formed chunks together. For the most part, this is accomplished by starch, but functional proteins may be involved, especially in production of texturized soy protein products. The formula must contain enough cereal grain and its starchy fractions for the dextrinized/gelatinized starch to cement product pieces together. Meat and fish meals and corn gluten do not contribute to cohesiveness. Oil and fat generally are detrimental to puffing of starch and texturization (striation) of soy protein at levels over 6%. If obtaining a porous and expanded product is not important, as much as 15–18% fat can be included in products made on single-screw extruders and 20–22% fat for twin-screw extruders.

In addition to the interference of fat in product expansion, natural and added antioxidants are partially lost by volatilization during flash-off of expanding products as they exit the die. Vitamins are often destroyed during extrusion processing, despite efforts to protect them by encapsulation. Lactose in dry milk and whey products will brown during heat processing in a reaction known as Maillard Browning. The preferred practice is to include minerals and heat-stable color materials in the extruder feedstock mix to ensure uniform distribution, but to hold back the fat, antioxidants, vitamins, milk/whey ingredients and flavoring materials for enrobing onto the formed product after extrusion and drying.

291 What are the rules of thumb regarding size reduction of ingredients?

Some extruders are capable of grinding whole corn and soybeans, but many people feel that these operations can be accomplished more efficiently with dedicated equipment like roller or hammer mills. A general rule of thumb is that the extruder feedstock does not contain any particles larger in size than one-third the diameter of the die holes. Passing the ingredients or extruder feedstock mixture through a positive classification device like a powered sieve or screen, or a hammer mill equipped with suitable-size screen, is recommended before extrusion. Equipment also should be protected by magnets and metal detectors to avoid damage from tramp metal.

292 What are the rules of thumb regarding mixing of ingredients for extrusion processing?

Various mixers are available for the dry feedstock. The minimum acceptable mixing type can be determined by stopping the mixer and taking samples randomly from the contents. The samples are next analyzed for an ingredient (like salt or a mineral component). Minimum acceptable mixing is the time required to reach the lowest stable standard deviation of analysis of the sample groups. The reader is referred to a statistics handbook for further instructions. Unfortunately, ingredients may 'unmix' under certain conditions. For example, heavy mineral components may settle to the bottom of the trough of a slow-moving conveyor. Dropping the mixture through a conveyor spout into a cone-formation pile may result in the heavier particles settling on the outside of the cone and especially at the bottom of the cone. In some instances, materials may be wetted and worked into dough before being fed to the extruder.

293 What are the most common problems that occur during extrusion?

Sometimes during processing an extruded product does not comply with the specifications set. The following are the most commonly observed problems in an extrusion system.

- Production rates are too low.
- The product is 'wedging'.
- There is too much surging.
- Product bulk density is too high.
- Product bulk density is too low.
- Product is distorted in shape.
- There are too many fines during processing.
- Product color is too dark.
- Product color is too light.
- Product breaks immediately after extruding.
- Moisture variation is too wide within the product.
- Moisture distribution is uneven.
- Extruder parts wear too quickly.
- Production palatability is low.

294 Why do these most common problems happen?

Everything happens for a reason and trouble-shooting is simply a matter of identifying a cause and then fixing it. The first step to take when identifying a problem is to gather accurate information about the raw material, hardware, processing conditions and final product. A problem occurs for one of the following reasons:

- The problem was always there and never was identified as a problem.
- Something has changed gradually with time and no one paid any attention until it became a major problem.
- Something happened suddenly.

295 How can we overcome the most common problems?

Here is a list of check points, which can help to solve most of the above-mentioned problems:

- Check the water supply to the conditioning cylinder and extruder barrel.
- Check the steam supply to the conditioning cylinder and extruder barrel.
- Check the indirect heating/cooling system of the extruder barrel.
- Check all parts for wear especially screws and shear locks.
- Check the shaft for wear or dents.
- Confirm that you are using the correct configuration for the product.
- Calibrate the dry feed rate on a regular basis.
- Confirm all your processing parameters.
- Check for die wear and make sure all the holes are open.
- Check knife sharpness and knife clearance against the die.

296 What causes surging during dry extrusion?

Surging is when the extruder runs empty and then full just before the die. Products produced by surging are either low or high in density. This is the most common problem when an extruder is not running full capacity. If this is the case, slow down the extruder shaft rpm and consider increasing the feed rate, or plugging some of the open die area. With time, extruder wear increases resulting in back flow of the material, which has a direct effect on surging. All the worn parts should be replaced in order to reduce surging.

Surging is a major problem when a recipe is high in fat or moisture, because of slippage and back flow. In this case there are two options: (1) reduce the oil or moisture content or (2) reconfigure the extruder with more aggressive and positive conveying screws. Too fine grind is always a problem for conveying. Surging will occur if the material is ground too fine, since the material does not convey well. To solve this problem, the grind needs to be a little coarser, since it flows better.

Remember that material that is too coarse is harder to cook and a product with coarser particles is more susceptible to breakage.

Extruder screws usually have a different taper or leading and trailing edges, affecting conveying and surging. Screws need to be indexed so the flights do not interfere and cause a problem like surging.

297 What causes wedging during extrusion?

Wedging is when the product flows unevenly from the die and is therefore cut thicker on one side than the other, One possible cause of this problem is that the product is being held back on one side by the knives as each knife blade travels over the die. This problem can be fixed by reducing the knife blade thickness or increasing the knife speed by reducing the number of knives. Adjusting the blade angle or grind will also help. Also make sure that the knife does not lie flat against the die face.

Another possible reason for this problem may be a product that flows unevenly from the die, being cut thicker on one side than the other. In a single screw extruder, velocity at the outside of the die tends to be higher than velocity at the center. This radial velocity gradient can be reduced by inserting a screen, a back-up (primary) die system before the final die or creating a dead space between the end of the screw and the die. Another way to handle this problem is by increasing the feed rate, or by increasing the extrudate viscosity by lowering oil or water addition. Product flow is affected by die design. Increasing the radius of tight corners will increase flow and shortening the land length on the side with slowest flow will also help.

298 What causes a lower feed rate in extrusion?

'Suddenly the feed rate needs to be adjusted because the extruder is filling up'. This situation usually occurs when screw parts are worn and cause back flow of the material. The extruder fills up at a fixed die pressure, reaching the point at which extrudate backs up to the inlet. The feed rate needs to be adjusted until the worn parts are replaced. This situation can be corrected temporarily by increasing the extruder rpm. Cooling the extruder barrel at the feeding zone has a positive effect on feeding and will be helpful. To make a decision whether or not to replace the worn parts, always measure the clearance between the screw and liner. When this clearance has doubled, it is time to replace the worn parts. It also depends upon the product which is being extruded. Some products have less effect on wear rates than others.

Another possible reason for a sudden lower feed rate may be an increase in the viscosity of the extrudate, which causes back flow. This problem can be solved by reducing the oil or water addition in the recipe. Some time variation in the addition of raw material like fresh meats can increase the viscosity by having more fat or water, which ultimately contributes to lowering the production rate.

Sometimes steam injection is too close to the feeding section and steam flow causes a feedback problem. Be sure that steam is being injected at the proper place and distribution is equal in the barrel. Sometimes steam can act as a 'pressure lock' and prevent dry feed from being conveyed down to the extruder. Always clean the screw before the starting an extruder. Sometimes old feed remains in the liners and can reduce the feed rate.

A misaligned or bent extruder shaft can cause a 'shearlock effect', causing feed to back up. Sometimes the screws' configuration can be back to front. Be sure all the screws are in the correct configuration and none have been put in back to front. Also check all the shearlocks for correct configuration and angles. It is recommended to start with the smallest diameter close to feed inlet. Be sure the grind is the same and the source of ingredients has not changed. Sometimes newly harvested ingredients are more difficult to grind than old ingredients.

299 What causes variation in product density during extrusion?

'Suddenly product density is too low or too high.' Density is a function of expansion, size and shape. This situation could occur for several reasons:

- Dies are worn out.
- Coarse grind.
- Oil content of dry feed is too high.
- Change in feed rate.

This problem can be solved by:

- Checking worn dies.
- Increasing feed rate.
- Checking for worn screw and liner.
- Increasing the conditioning temperature.
- Increasing or decreasing the water in the preconditioner.
- Adjusting the feed rate.
- Increasing the steam injection in the extruder barrel.
- Cooling the barrel.
- Increasing the starch content in the recipe.

Always check product specification for size and shape. A minor change in specification can have a major effect on the product density. If dies are worn, the extruder operator may be controlling size by reducing expansion ratios, thus increasing bulk density.

300 What causes product to break or become distorted?

'A product can get distorted, if it is too soft, or it is under cooked. Sometimes products can be damaged by knives and conveying systems.'

This problem can be solved by changing the following parameters:

- Make sure knife speed is not too high.
- Increase number of blades.
- Check for plugged dies and increase the number of holes.
- Check knife hoods for proper installation, sometimes product can hit the hood and break.
- Check product for excessive oil or water.
- Reduce air velocity for a negative air system.
- Reduce knife width and angle of attack of knife blades.
- Small pieces of ingredients in the final product may act as a breaking point.

301 What causes color variation during extrusion?

'Suddenly products vary in color.' Color is usually a function of degree of cook, particle size of the ingredients and added color. Coarsely ground ingredients make products look darker. Specification and levels of addition of color also affect the final product color.

Color variation can be controlled by changing the following parameters:

- Check grind and make sure the size has not changed.
- Check variation of rate of dry feed and addition of color.
- Blend the color evenly in the premix.
- Check feed formulation changes.
- Natural color can vary more than artificial colors.
- Some natural color varies in their intensity, i.e. iron oxide.

302 How can problems in picking up the product by the screw be reduced?

The difficulty of wetting dry powders to enhance feedstock traction (pickup by the screw) at the inlet of the extruder has been mentioned. This is less of a problem in twin screw extruders than in single screw machines, but still can be significant. Preconditioning enhances screw pickup by wetting the product. Problems exist in picking up dough and sticky products and can be reduced by: (1) building a sloped feed throat or undercut feed throat hopper in the feed end barrel which keeps the dough from riding on top of the revolving screw; (2) building a packer or second short, powered parallel screw into the feed throat; (3) building a multi-screw live-bottom feeder at the inlet; or (4) using a square-throat positive displacement pump to feed the extruder, provided the feedstock is pumpable.

Generally, a plug of product must be maintained between the feed and the cooking sections to prevent blowback of steam. This can be done by cooling the transition section with a cold water jacket often in conjunction with an appropriately sized shear lock.

Index

£230.00